A-LEVEL YEAR 2

STUDENT GUIDE

AQA

Biology

Topics 5 and 6

Energy transfers in and between organisms

Organisms respond to changes in their internal and external environments

Pauline Lowrie

Philip Allan, an imprint of Hodder Education, an Hachette UK company, Blenheim Court, George Street, Banbury, Oxfordshire OX16 5BH

Orders

Bookpoint Ltd, 130 Park Drive, Milton Park, Abingdon, Oxfordshire OX14 4SE

tel: 01235 827827

fax: 01235 400401

e-mail: education@bookpoint.co.uk

Lines are open 9.00 a.m.–5.00 p.m., Monday to Saturday, with a 24-hour message answering service. You can also order through the Hodder Education website: www.hoddereducation. co.uk

© Pauline Lowrie 2016

ISBN 978-1-4718-5669-3

First printed 2016

Impression number 5 4 3 2 1

Year 2020 2019 2018 2017 2016

This guide has been written specifically to support students preparing for the AQA A-level Biology examinations. The content has been neither approved nor endorsed by AQA and remains the sole responsibility of the author.

Cover photo: Sergey Nivens/Fotolia

Typeset by Integra Software Services Pvt. Ltd, Pondicherry, India

Printed in Italy

Hachette UK's policy is to use papers that are natural, renewable and recyclable products and made from wood grown in sustainable forests. The logging and manufacturing processes are expected to conform to the environmental regulations of the country of origin.

Contents

Content Guidance

Questions & Answers

■ Getting the most from this book

Exam tips
Advice on key points in the text to help you learn and recall content, avoid pitfalls, and polish your exam technique in order to boost your grade.

Knowledge check
Rapid-fire questions throughout the Content Guidance section to check your understanding.

Knowledge check answers
1 Turn to the back of the book for the Knowledge check answers.

Summaries
- Each core topic is rounded off by a bullet-list summary for quick-check reference of what you need to know.

Exam-style questions

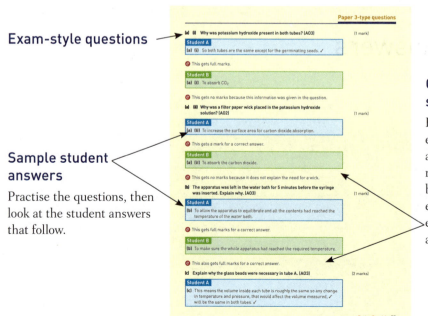

Sample student answers

Practise the questions, then look at the student answers that follow.

Commentary on sample student answers

Find out how many marks each answer would be awarded in the exam and then read the comments (preceded by the icon **e**) following each student answer showing exactly how and where marks are gained or lost.

■ About this book

This guide will help you to prepare for AQA A-level Biology topics 5 and 6. These topics are examined in paper 2 (together with topics 7 and 8) and in paper 3 (together with topics 1–4 and 7–9).

The **Content Guidance** covers all the facts you need to know and concepts you need to understand for topics 5 and 6. It is important to focus on *understanding* and not just learning facts, as the examiners will be testing your ability to apply what you have learned in new contexts. This is impossible to do unless you really understand everything. The Content Guidance also includes exam tips and knowledge checks to help you prepare for your exams.

The **Questions & Answers** section shows you the types of questions you can expect in the exam. It would be impossible to give examples of every kind of question in one book, but these should give you a flavour of what to expect. Two students, student A and student B, have attempted each question. Their answers, along with the accompanying comments, should help you to see what you need to do to score a good mark — and how you can easily *not* score a mark even though you probably understand the biology.

What can I assume about the guide?

You can assume that:
- the basic facts you need to know and understand are stated explicitly
- the major concepts you need to understand are explained clearly
- the questions at the end of the guide are similar in style to those that will appear in the final examination
- the questions assess the different assessment objectives
- the standard of the marking is broadly equivalent to that which will be applied to your answers

How should I use this guide?

The guide lends itself to a number of uses throughout your course — it is not *just* a revision aid. You could:
- use it to check that your notes cover the material required by the specification
- use it to identify your strengths and weaknesses
- use it as a reference for homework and internal tests
- use it during your revision to prepare 'bite-sized' chunks of related material, rather than being faced with a file full of notes

You could use the Questions & Answers section to:
- identify the terms used by examiners and show what they expect of you
- familiarise yourself with the style of questions you can expect
- identify the ways in which students gain, or fail to gain, marks

Develop your examination strategy

Just as reading the *Highway Code* alone will not help you to pass your driving test, this guide cannot help to make you a good examination candidate unless you develop and maintain all the skills that examiners will test in the final exams. You also need to be aware of the types of questions examiners ask and where to find them in the exams. You can then develop your personal examination strategy. But be warned, this is a highly personal and long-term process — you cannot do it a few days before the exam.

Things you *must* do

- Clearly, you must know some biology. If you don't, you cannot expect to get a good grade. This guide provides a succinct summary of the biology you must know.
- Be aware of the skills that examiners *must* test in the exams. These are called assessment objectives and are described in the AQA Biology specification.
- Understand the weighting of the assessment objectives that will be used in the exam. These are as follows:

Assessment objective	Brief summary	Marks in A-level paper 1/%	Marks in A-level paper 2/%	Marks in A-level paper 3/%
AO1	Knowledge and understanding	44–48	23–27	28–32
AO2	Application of knowledge and understanding	30–34	52–56	35–39
AO3	Analyse, interpret and evaluate scientific information, ideas and evidence	20–24	19–23	31–35

- Use past questions and other exercises to develop all the skills that examiners must test. Once you have developed them all, keep practising to maintain them.
- Understand where in your exams different types of questions occur. For example, the final question on A-level paper 3 will always be an essay worth 15 marks, testing mainly AO1. If that is the skill in which you feel most comfortable, why not attempt this question first?
- Remember that mathematical skills account for about 10% of the marks. Make sure you can carry out these calculations, including percentages, ratios and rates of reaction. Also remember that at A-level you are required to understand some statistical tests.
- You need to be familiar with the techniques you have learned in the required practicals and be able to describe how these techniques might be used in a different context. Also, you need to be able to evaluate practical investigations and data presented to you in the exam. Answers to the questions set in the required practicals in this book are given on pages 83–84.

Content Guidance

■ Energy transfers in and between organisms

Photosynthesis

Photosynthesis is a two-stage process that takes place in the chloroplast. There is a light-dependent stage, which takes place in the grana of the chloroplast. In this stage, ATP and reduced NADP are made, which are used in the light-independent stage. The light-independent stage takes place in the stroma. This is summarised in Figure 1.

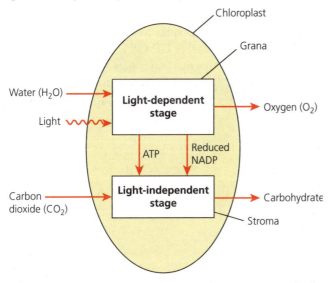

Figure 1 A summary of photosynthesis

Let's look at the light-dependent stage of photosynthesis in more detail (see Figure 2). Chlorophyll is a green pigment that is found embedded in the membranes of the grana. When light hits a chlorophyll molecule, this causes one of its electrons to absorb some of the light energy and move to a higher energy level. We say that the electron has become **excited**. This process is called **photoionisation**. The electron passes to a carrier protein in the membrane. From there, the electron passes through a series of protein carrier molecules in the membrane, called an **electron transfer chain**. As the electron moves down this electron transfer chain, energy is lost and an ATPase makes ATP from ADP and P_i.

The electron that has been lost by the chlorophyll molecule needs to be replaced. A process called **photolysis** occurs. Light energy splits water into oxygen (which is given off as a waste product), electrons, which replace those lost by chlorophyll, and

Knowledge check 1

The diagram shows a chloroplast. Which letter represents **a** the grana and which represents **b** the stroma?

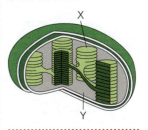

Exam tip

Remember that P_i is shorthand for inorganic phosphate.

H^+ ions (or **protons**). The H^+ ions join up with the electrons that have passed through the electron transfer chain to become hydrogen atoms, and they join with a coenzyme called NADP to become NADPH, which is a **reduced coenzyme**.

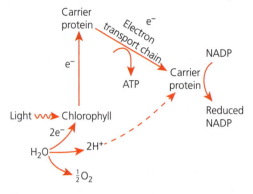

Figure 2 The light-dependent reaction

The next stage of photosynthesis is the **light-independent** stage. You can see this in Figure 3. Carbon dioxide joins up with a five-carbon molecule called ribulose bisphosphate (RuBP for short). The resulting molecule immediately splits to form two molecules of a three-carbon compound, glycerate-3-phosphate, or GP. GP is then reduced to triose phosphate (TP) by NADPH. The energy for this reaction comes from ATP. Triose phosphate can be used to regenerate RuBP (using ATP) but also to synthesise all the organic molecules the plant needs, such as cellulose, amino acids, glucose, etc.

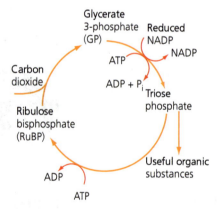

Figure 3 The light-independent stage of photosynthesis

Limiting factors

As we have just seen, photosynthesis relies on a number of factors. If any one of these factors is in short supply, it becomes a **limiting factor**. Only one factor can limit the rate of photosynthesis at one time. If that limiting factor is increased, the rate of photosynthesis increases.

Exam tip

Remember OILRIG — oxidation is loss, reduction is gain.

Knowledge check 2

Although the light-independent stage of photosynthesis doesn't use light energy, it does not take place in the dark. Explain why.

Knowledge check 3

Explain why we can say GP is reduced when it is converted to TP.

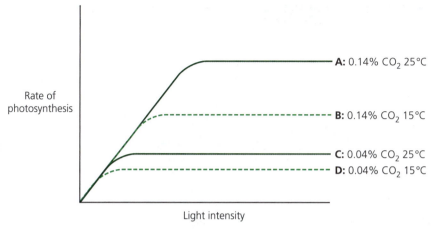

Rate of photosynthesis

A: 0.14% CO_2 25°C

B: 0.14% CO_2 15°C

C: 0.04% CO_2 25°C

D: 0.04% CO_2 15°C

Light intensity

Figure 4 The effect of different factors on the rate of photosynthesis

Figure 4 shows the rate of photosynthesis under different conditions. You can tell that light intensity is the limiting factor in the part of the graph that slopes upwards, since increasing light intensity increases the rate of photosynthesis. However, when the graphs flatten off, another factor is limiting photosynthesis. For example, the limiting factor in D must be temperature. You can tell this by comparing C and D. C has a higher rate of photosynthesis than D, but the only factor that has changed is that the temperature is higher. So we know that is the limiting factor.

Knowledge check 4

a What is limiting the rate of photosynthesis in the flat part of graphs C and B in Figure 4?

b Suggest what is limiting photosynthesis in the flat part of graph A.

Exam tip

You may be given data to interpret about the effects of some of these factors on crop yields. You may be asked to carry out calculations and to make judgements based on the data.

Required practical 7

Chromatography of plant pigments

A student collected some leaves from a plant and then ground them up with some sand and a solvent in a pestle and mortar. The student marked an origin line in pencil near the bottom of a piece of chromatography paper. He placed a spot of the solvent containing the pigments from the leaf on the origin line and dried the spot using a hairdryer. He repeated this ten times, then placed the chromatography paper in a boiling tube containing a suitable solvent. When the solvent had almost reached the top of the paper, he marked this line — the solvent front — using a pencil. He dried the paper. The finished chromatogram is shown in Figure 5.

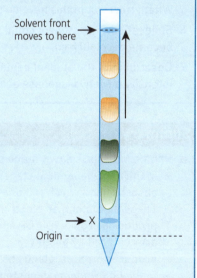

Solvent front moves to here

X

Origin

Figure 5 Chromatogram

Questions

1 Why was the origin line drawn in pencil?

2 Why did the student need to grind up the leaves with the solvent?

3 Why did the student apply a spot of solvent containing pigment to the origin line several times, drying in between?

4 Why was the origin line above the level of solvent in the boiling tube?

5 The student put a bung in the boiling tube after leaving the chromatography paper in it. Explain why.

6 Calculate the R_f value of spot X.

7 Suggest an advantage to a plant of having several different pigments in the chloroplast, in addition to chlorophyll.

Required practical 8

Investigating dehydrogenase activity in chloroplasts

A chloroplast suspension was made by homogenising some fresh spinach leaves in isotonic saline. The mixture was filtered and centrifuged. The pellet, containing chloroplasts, was re-suspended in isotonic saline solution. A student set up three test tubes as follows:

Tube number	Contents of tube/cm³		
	Chloroplast suspension	Isotonic saline	DCPIP
1	1	0	10
2	0	1	10
3	1	10	0

DCPIP is a blue dye that accepts electrons from electron transfer chains. When it accepts electrons it becomes colourless.

The three tubes were set up in a test tube rack 15 cm from a bright bench lamp.

The student observed that tube 1 became decolourised after 7 minutes. No changes were observed in the other two tubes.

Questions

1 Explain why tube 1 became decolourised.

2 Explain the purpose of (a) tube 2, (b) tube 3.

3 Why was it important to re-suspend the chloroplasts in *isotonic* saline solution?

4 Before use, the chloroplast suspension was kept ice-cold. Explain why.

5 Sodium hydrogen carbonate could be used to increase the carbon dioxide concentration in the tubes. What effect would this have on the time taken for tube 1 to decolourise? Explain your answer.

6 Suggest how the student could have modified this investigation to measure the rate at which tube 1 became decolourised.

7 Describe how the student could modify this investigation to measure the effect of changing light intensity on dehydrogenase activity in chloroplasts.

Summary

- Photosynthesis is a two-stage process.
  - In the light-dependent stage, chlorophyll absorbs light, which releases electrons. These pass along an electron transfer chain producing reduced NADP and ATP. Electrons are replaced by photolysis of water, which produces protons, electrons and oxygen.
  - In the light-independent stage, carbon dioxide reacts with ribulose bisphosphate (RuBP) to produce two molecules of GP. ATP and reduced

NADP from the light-dependent stage are used to reduce GP to triose phosphate (TP). TP can be used to regenerate RuBP and the rest is converted to useful organic substances.

- The rate of photosynthesis can be limited by factors such as the availability of light, carbon dioxide and temperature. In agriculture, farmers may try to overcome the effect of these limiting factors to increase yields.

Respiration

Respiration takes place in all living cells. It is a process in which organic molecules (called respiratory substrates) are broken down in stages to release ATP. The stages in respiration are summarised in Figure 6.

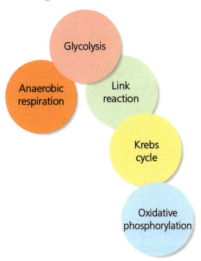

Figure 6 The stages in respiration

The first stage, summarised in Figure 7, is **glycolysis**. This occurs in the cytoplasm.

■ Glucose is not very reactive, so at first glucose is **phosphorylated** to glucose phosphate, using ATP. (Remember that phosphorylation means adding a phosphate group.)

■ Glucose phosphate is broken down into two molecules of triose phosphate.

■ Triose phosphate is oxidised to form pyruvate, yielding reduced NAD and ATP.

Notice that no oxygen is required for glycolysis, so this stage takes place whether or not oxygen is present. However, when oxygen is available, pyruvate enters the link reaction, which takes place in the matrix of the mitochondrion. This is shown in Figure 8.

In the link reaction, the three-carbon molecule, pyruvate, is converted to a two-carbon molecule, acetate. This involves the production of reduced NAD and the carbon is released as carbon dioxide. Acetate then combines with coenzyme A to form acetyl coenzyme A.

Knowledge check 5

Is the conversion of pyruvate to acetate an oxidation reaction or a reduction reaction? Explain your answer.

Acetyl coenzyme A then enters the Krebs cycle. This also takes place in the matrix of the mitochondrion. This process is summarised in Figure 9.

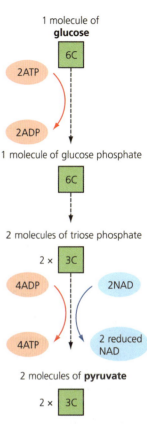

Figure 7 Glycolysis

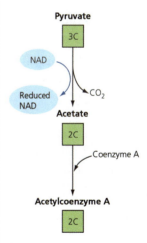

Figure 8 The link reaction

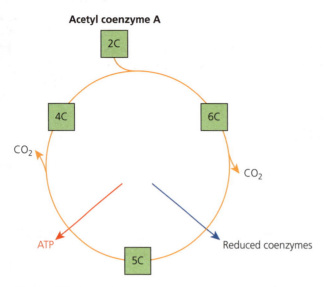

Figure 9 Krebs cycle

Acetyl coenzyme A (2C) joins with a 4C compound to produce a 6C compound. This is broken down to a 5C compound, which in turn is broken down into a 4C compound. This combines with another molecule of acetyl coenzyme A and the cycle continues. You will notice that the acetylcoenzyme A (all that remains of the glucose we started with) is completely broken down here. Two carbon atoms enter the cycle and both are lost in the form of carbon dioxide. In addition, reduced coenzymes are made and one ATP is made for each turn of the cycle.

The ATP made in the Krebs cycle, and in glycolysis, is made by **substrate-level phosphorylation**. This is because the ATP is made when one substrate is converted to another.

A lot of reduced coenzymes have been made in glycolysis, the link reaction and the Krebs cycle. These enter the electron transfer chain, in the inner membrane of the mitochondrion. You can see this in Figure 10.

You will see that the reduced coenzyme passes its hydrogen to a carrier protein in the electron transfer chain. This is split into a proton (or hydrogen ion, H^+) and an electron. The protons pass through to the space between the inner and the outer mitochondrial membrane, while the electrons pass through the proteins in the electron transfer chain. The protons return back through the ATP synthase enzyme in the membrane, producing ATP. The protons and electrons recombine to make hydrogen atoms again, and they combine with oxygen to make water. Oxygen is the last carrier in the electron transfer chain. Making ATP using an electron transfer chain is called **oxidative phosphorylation**. The electron transfer chain in the light-dependent stage of photosynthesis is very similar.

Cells can use other respiratory substrates, such as lipids and amino acids, which can enter the Krebs cycle.

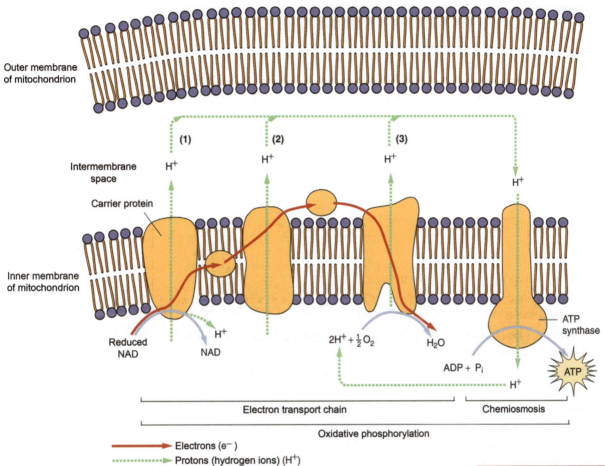

Figure 10 The electron transfer chain

The processes described so far are the stages of **aerobic respiration**. If oxygen is not present, electron transfer cannot happen; nor can the Krebs cycle or the link reaction. Therefore, in anaerobic respiration pyruvate formed at the end of the glycolysis is converted to another product (see Figure 11). This is necessary to recycle NAD.

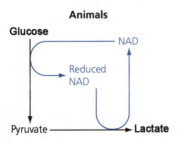

Figure 11 Anaerobic respiration

Knowledge check 6

What would happen in anaerobic conditions if pyruvate was not converted to lactate in animals and some bacteria, or to ethanol in plants and most microorganisms?

Knowledge check 7

Explain why the yield of ATP per glucose is much higher when aerobic, rather than anaerobic, respiration occurs.

Required practical 9

Investigating respiration in yeast

A student set up five tubes, each containing $2\,cm^3$ of a yeast suspension and $2\,cm^3$ of a substrate. The tubes were placed in a water bath at 30°C and the volume of gas given off in 40 minutes was measured. The results are shown in the table.

Contents of tube	Volume of gas given off in 40 minutes/cm³
Yeast + glucose	13.6
Yeast + sucrose	10.3
Yeast + lactose	4.7
Yeast + glycerol	6.4
Yeast + distilled water	3.5

Questions

1 Name the gas being given off by the yeast.

2 When distilled water, containing no respiratory substrate, was added, there was still some gas given off. Suggest why.

3 When the yeast has glucose added, more gas is given off than when sucrose is added. Suggest why.

4 Less gas is given off when the yeast has lactose or glycerol added than when sucrose is added. Suggest an explanation for this.

5 Another student said that there should be a control investigation, in which five more tubes are set up in the same way, but with boiled yeast suspension. Do you agree? Give a reason for your answer.

Summary

- Aerobic respiration consists of four stages: glycolysis, the link reaction, the Krebs cycle and electron transfer.
  - In glycolysis, glucose is broken down to two molecules of pyruvate, producing a small net gain of ATP and reduced coenzymes. This takes place in the cytoplasm.
  - In the link reaction in the mitochondrial matrix, pyruvate is converted to acetyl coenzyme A, producing reduced coenzymes and carbon dioxide.
  - In the Krebs cycle in the mitochondrial matrix, acetyl coenzyme A is completely broken down, yielding carbon dioxide, ATP and reduced coenzymes.
  - In electron transfer, on the inner mitochondrial membrane, hydrogen atoms from the reduced coenzymes split into protons and electrons. The electrons pass through the electron transfer chain and the protons pass through an ATP synthase molecule, producing ATP. Oxygen is the final electron acceptor.
- Anaerobic respiration consists of glycolysis only. However, pyruvate is converted into lactate (in animals and some bacteria) or to ethanol and carbon dioxide (in plants and most microorganisms) to oxidise the reduced coenzymes, allowing glycolysis to continue.

Energy and ecosystems

The basis of almost all ecosystems is green plants. These absorb light energy and carbon dioxide and produce organic matter in photosynthesis. As we have just seen, a lot of the TP produced in photosynthesis is used to make sugars that are used in respiration. However, the rest can be used to make other molecules, such as starch, proteins, lipids and cellulose, which make up the biomass of the plants.

Biomass is the total mass of organisms in a given area. It is measured in terms of carbon or dry mass of tissue per given area per given time. The dry mass is used, rather than the fresh mass, because the water content of organisms can vary, depending on environmental conditions.

Exam tip

The dry mass of grass in a field would be found by sampling a square metre at random. The grass from the area would be removed, including the roots, and soil would be washed off. The grass would be placed in an oven at 100°C (you can remember this as the temperature at which water vaporises). After a few hours, the grass is weighed, then returned to the oven for a further hour. It is weighed again and the process repeated until two weights are the same. This is called weighing to constant mass.

The energy stored in the biomass can be found using a calorimeter, shown in Figure 12.

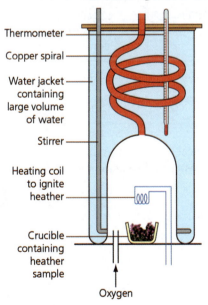

Thermometer
Copper spiral
Water jacket containing large volume of water
Stirrer
Heating coil to ignite heather
Crucible containing heather sample
Oxygen

Figure 12 A calorimeter

The calorimeter contains a known volume of water. A known dry mass of biological material is burned inside the calorimeter. The apparatus is well insulated and designed so that as much of the heat energy as possible is transferred to the water. We know that it takes 4.2 joules of heat energy to raise the temperature of 1 cm^3 of water by 1°C, so we can work out the energy in the sample.

Knowledge check 8

Use the data in the table to find the energy content of 1 g of willow.

Mass of willow/g	0.5
Volume of water in calorimeter/cm^3	650
Temperature of water at start/°C	19
Temperature of water at end/°C	22

Gross primary production (GPP) is the chemical energy stored in plant biomass. We usually measure this for a specific area in a given time. However, some of the compounds produced in photosynthesis are metabolised in respiration. Therefore **net primary production (NPP)** is the chemical energy stored in plant biomass after respiratory losses have been taken into account. NPP is the chemical energy store in plants that is available to primary consumers and decomposers.

NPP = GPP − R

where R = respiratory losses to the environment.

Knowledge check 9

Give suitable units for **a** biomass of grass in a meadow, **b** gross primary production, **c** net primary production.

Knowledge check 10

Complete the table.

Ecosystem	NPP/kJ m^{-2} y^{-1}	GPP/kJ m^{-2} y^{-1}	R/kJ m^{-2} y^{-1}
Mature rainforest (Puerto Rico)	?	189 000	134 400
Lucerne (alfalfa) crop (USA)	63 840	?	38 640

In a similar way, we can calculate the net production of consumers (N) as

N = I − F − R

where I = chemical energy in ingested food, F = chemical energy lost to the environment in faeces and urine, and R = respiratory losses to the environment.

Knowledge check 11

a Use the data in the table to calculate the net production of a cow.

Energy in food eaten	3050
Heat lost in respiration	1025
Energy lost in faeces and urine	1900

b Suggest suitable units for the figures in the table.

Knowledge check 12

The net primary production of the grass in the field where the cow in knowledge check 11 was grazing was 21 250. Apart from energy lost by the cow in respiration, faeces and urine, give *one* reason for the cow's net production being so much less than the net primary production.

In agriculture, farmers may increase the efficiency of energy transfer by adopting various farming practices, such as rearing livestock indoors, to reduce energy losses in heat and movement. Intensively reared cattle may be fed concentrates that are more digestible than grass, reducing energy lost in faeces.

Figure 13 shows how energy is transferred through a food chain. At every trophic level, energy is lost as heat in respiration.

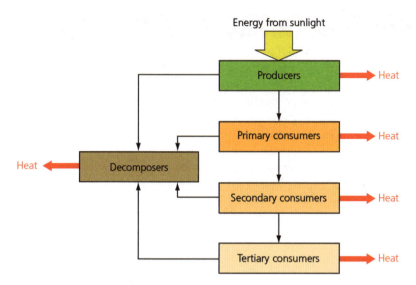

Figure 13 How energy is transferred through a food chain

Figure 14 shows that only a small amount of energy is transferred from one trophic level to the next. The most inefficient stage is converting the energy in sunlight to biomass in the producers. This is because:

- some of the light is the wrong wavelength for photosynthesis
- some of the sunlight does not hit a chloroplast
- some of the light energy is converted to heat energy
- some of the light energy is reflected from the leaf or transmitted through it

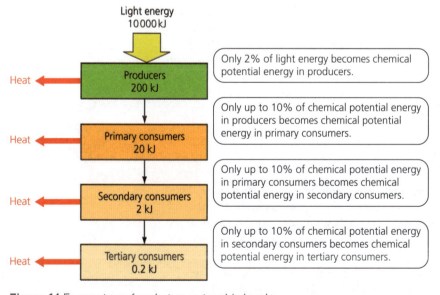

Figure 14 Energy transfers between trophic levels

Knowledge check 13

Why do food chains rarely have more than five trophic levels?

Knowledge check 14

Why is so little energy transferred from one trophic level to the next?

Summary

- Most of the sugars made by green plants in photosynthesis are used by the plant in respiration. The rest of the sugars are used to make organic compounds that form the biomass of plants. Biomass is measured as mass of carbon or dry mass of tissue per unit area per unit time. The chemical energy in biomass can be measured using a colorimeter.
- Gross primary production (GPP) is the chemical energy stored in plant biomass in a given area or volume in a given time.
- Net primary production (NPP) is the chemical energy stored in plant biomass after respiratory losses have been taken into account. Net primary production is the chemical energy available to the next trophic level.
- In agriculture, different farming practices may be used to increase the efficiency of energy transfer and improve productivity.

Nutrient cycles

Decomposers

Saprobionts are organisms — mainly fungi and bacteria — that secrete enzymes on to dead organic remains of organisms. They digest this material and absorb some of the nutrients from it. However, some of the nutrients remain in the surroundings and may be absorbed by other organisms. Some of the nutrients they release are mineral ions such as phosphate and nitrate, which are very important in ecosystems. If decomposers were not recycling these nutrients, dead plant and animal material would remain undecomposed for long periods of time and ecosystems would have much lower productivity.

Mycorrhizae are fungi that are associated with the roots of most plants. They grow in and around the plants and are often specific to the plant. Mycorrhizae, like most other fungi, are made up of tiny threads called hyphae, which increase the surface area of the roots for taking up water and nutrients from the soil. The mycorrhizal fungi secrete enzymes that digest dead organic matter in the soil, then they absorb the nutrients released and transport them back into the plant roots. In return, the fungi obtain carbohydrates from the plant. This is called a **mutualistic** relationship, since both the plant and the mycorrhizae benefit.

The phosphorus cycle

The phosphorus cycle is shown in Figure 15.

You will see that producers absorb phosphate ions from the soil into their roots. These phosphate ions are incorporated into organic molecules in plant biomass. Primary consumers obtain phosphorus-containing compounds when they eat the plants and these become incorporated into organic compounds in the primary consumer. Similarly, secondary consumers obtain phosphorus in the biomass of primary consumers. When these organisms die, or produce faeces and urine, decomposers recycle the organic material in the manner already described, making phosphates available to plants again via the roots.

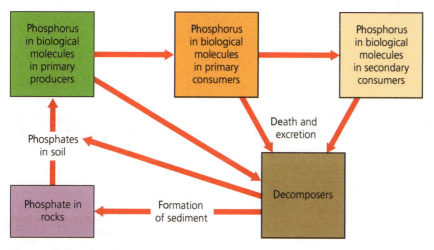

Figure 15 The phosphorus cycle

The nitrogen cycle

Figure 16 shows the nitrogen cycle.

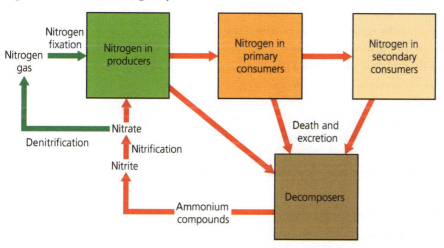

Figure 16 The nitrogen cycle

This cycle is similar to the phosphorus cycle.

- Ammonifying bacteria release ammonium ions from nitrogen-containing compounds such as proteins and urea in dead organic material and waste.
- Nitrifying bacteria convert ammonium ions to nitrite and then nitrate ions.
- Nitrogen-fixing bacteria may be free-living in the soil or in root nodules of leguminous plants such as peas, beans and clover. These produce ammonium ions from atmospheric nitrogen.
- Denitrifying bacteria, in waterlogged, anaerobic soils, convert ammonium and nitrate ions back into nitrogen gas and oxides of nitrogen.

Knowledge check 15

Name two organic compounds found in living organisms that contain phosphorus.

Exam tip

Make sure you understand the role of bacteria in the processes outlined above. This is an important part of the nitrogen cycle.

Knowledge check 16

Farmers sometimes plant a crop of 'green manure'. This is a crop such as clover or vetch that has nitrogen-fixing bacteria in its root nodules. The plant is left to grow for a while, then ploughed back into the soil. Suggest *two* advantages to a farmer of growing green manure.

Use of fertilisers

Fertilisers are added to soils to add nutrients, especially nitrogen, phosphorus and potassium (NPK). Fertilisers may be natural, e.g. farmyard manure, or artificial.

Natural fertilisers:

- have a variable composition, but will contain trace elements in addition to N, P and K, which may be beneficial to the crop
- add organic matter to the soil, which improves soil quality
- release nutrients slowly over a long period of time, reducing the risk of leaching

If the manure is produced on the same farm, it is cheap to use and gets rid of waste that might otherwise cause pollution. However, manure is heavy to transport and can cause soil compaction when heavy machinery is used.

Artificial fertilisers:

- have a known composition so can be applied in the correct concentration
- are concentrated so can be applied in smaller amounts, reducing the damage done by heavy machinery
- release nutrients more quickly and are more likely to cause leaching

Ions such as phosphates and nitrates are soluble and can be washed out of the soil into streams and rivers. We call this **leaching**. When these nutrients are added to water courses such as rivers, this can lead to **eutrophication**.

- The nutrients increase the growth of aquatic plants and algae.
- These grow on the surface, blocking sunlight from the plants that grow lower down. This is called an **algal bloom**.
- The plants beneath die and are decomposed by saprobionts.
- These saprobionts respire using oxygen, so there is little or no oxygen left in the water.
- As a result, aquatic organisms such as fish and invertebrates die.

This is summarised in Figure 17.

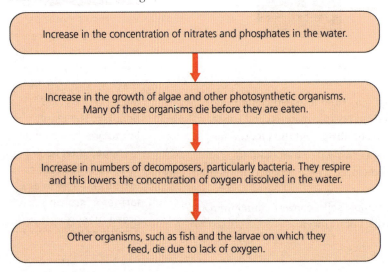

Figure 17 Eutrophication

Knowledge check 17

Why is it necessary for farmers to apply fertilisers?

Knowledge check 18

Farmers also add fertilisers to fields of grass used for animal grazing. Explain why.

Summary

- Nutrients are recycled within ecosystems. This includes the nitrogen and phosphorus cycles. Microorganisms are important in the recycling of nutrients.
- Saprobionts are microorganisms that digest dead organic matter into smaller, soluble molecules that can be absorbed by the roots of plants. Mycorrhizae are fungi associated with plant roots. These increase the surface area of roots for the uptake of water and mineral ions.

- In the nitrogen cycle, bacteria are involved in the processes of ammonification, nitrification, nitrogen fixation and denitrification.
- Farmers apply fertilisers to soil to replace the nitrates and phosphates lost by harvesting crops and removing livestock. These fertilisers may be natural or artificial. Using fertilisers may lead to environmental problems, including leaching and eutrophication.

■ Organisms respond to changes in their internal and external environments

Stimuli, both internal and external, are detected and lead to a response

Survival and response

Tropisms

A **tropism** is a growth response to a directional stimulus. This is seen in flowering plants, which will grow towards, away from or at an angle to a directional stimulus such as gravity or sunlight. This results from a growth-regulating chemical in plants, called indoleacetic acid (IAA) or auxin. Plant shoots are known to grow towards light, which is called **positive phototropism**. Many investigations were carried out to explain this response, some of which are summarised in Figure 18. These investigations used coleoptiles. These are shoots produced by germinating seedlings from the grass family. Coleoptiles are similar to shoots but much easier to study because they don't have leaves or buds.

From these investigations, it can be deduced that:
- the stimulus (in this case directional light) is perceived by the apex of the stem
- auxin is produced at the apex and diffuses down the stem to the region where the growth response occurs
- auxin is transported from the illuminated side to the shaded side. Here, it stimulates cell elongation so the stem bends towards the light

Figure 19 shows an investigation that demonstrates there is more auxin on the shaded side.

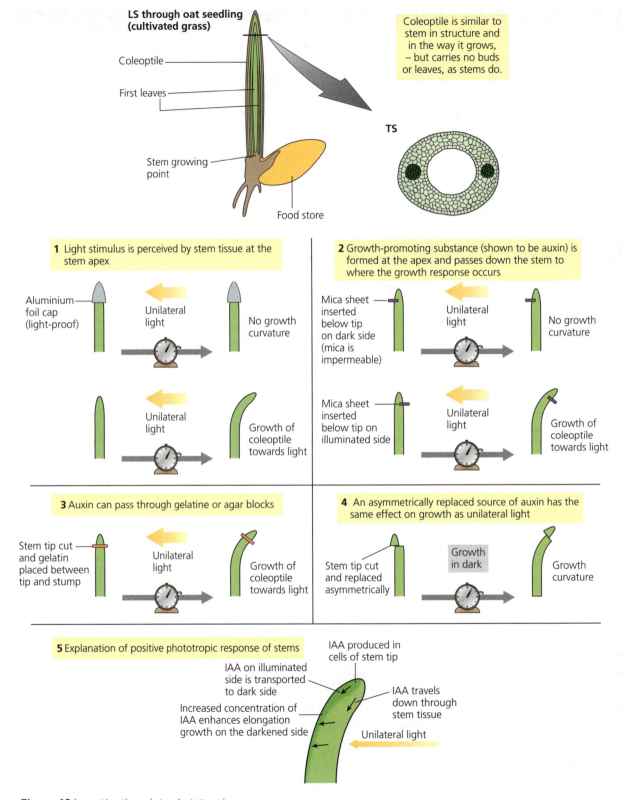

Figure 18 Investigations into phototropism

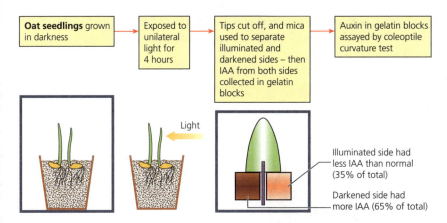

Figure 19 Investigating how auxin produces phototropism

Other investigations have shown that the more auxin that is present in the shoot, the more growth is stimulated, until a certain point when the auxin concentration becomes inhibitory. This is shown in Figure 20.

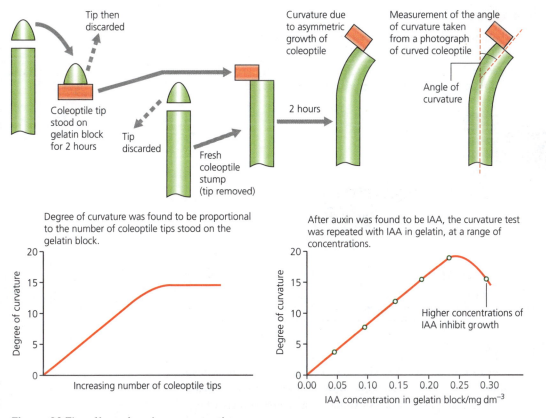

Figure 20 The effect of auxin concentration

Roots grow towards gravity. This is called positive geotropism. In plants, some of the cells contain starch-storage organelles called amyloplasts. These settle at the bottom of the cell under the force of gravity. Auxin (IAA) seems to be transported to the side of the cell where the amyloplasts are. In roots, a high concentration of auxin inhibits cell elongation. Therefore the lower side of the root elongates less than the upper side, causing the root to grow towards gravity. This is shown in Figure 21.

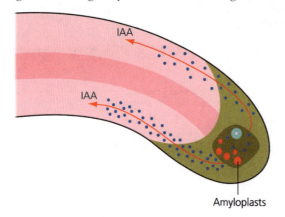

Figure 21 The mechanism of geotropism in the root

The diagram illustrates an investigation to show geotropism. In A, germinating seedlings are attached to a slowly rotating drum called a clinostat, so that the direction of gravity is constantly changing. In B, the germinating seedlings are kept in the same position throughout.

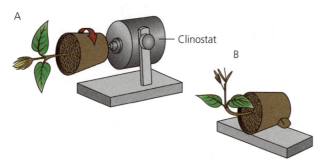

a Use your knowledge of IAA to explain the different responses to gravity of the shoots in A and B.

b Which investigation, A or B, is the control? Give a reason for your answer.

Taxes and kineses

Animals show responses to stimuli, too. A **taxis** is a directional movement in response to a directional stimulus. The organism may move towards, away from or at an angle to the stimulus. For example, simple flatworms exhibit positive chemotaxis when they move towards a food source, and sperm swim towards the ovum in the female reproductive tract by chemotaxis. Single-celled photosynthetic organisms move towards light by positive phototaxis. Many insects orientate their flight at an angle to the sun, allowing species such as honey bees to fly away from the nest to find nectar and then return to the nest. Some insects fly at 90° to a light source, which accounts for many moths flying around street lights at night.

Kinesis is another innate response in which the organism moves around at random. The rate of movement and turning is increased when conditions are unfavourable and decreases when conditions are more favourable. An example of a kinesis is in woodlice. Woodlice move and turn faster in light, dry conditions and slow down when they are in damp, dark conditions. Woodlice carry out gas exchange on gills under their body which need moist conditions. This kinesis behaviour ensures that woodlice find a suitable moist environment where they are out of sight of predators.

A student used a choice chamber like the one illustrated to investigate woodlice behaviour.

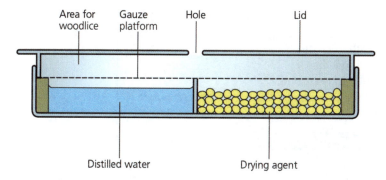

Half of the choice chamber was exposed to light and half was in darkness, so there were four areas as shown on the right.

Ten woodlice were placed in the choice chamber. Every minute for 10 minutes the student recorded the number of woodlice in each section. The table shows the results.

Time/minutes	Number of woodlice in each section			
	Light dry	Dark dry	Light damp	Dark damp
1	2	3	3	2
2	1	4	3	3
3	2	3	3	2
4	1	4	2	3
5	1	3	2	4
6	0	3	2	5
7	0	2	2	6
8	0	1	2	7
9	0	0	2	8
10	0	0	0	10
11	0	0	0	10

a What can you conclude from these results?
b Suggest how the student could confirm whether this behaviour is a taxis or a kinesis.

Required practical 10

Investigation into the effect of an environmental variable on the movement of an animal using either a choice chamber or a maze

This is just one example of an investigation that you could carry out.

A student wanted to find out whether woodlice show turn alternation behaviour. This is behaviour in which animals make a turn in one direction and then make the next turn in the opposite direction. The student set up a cardboard maze like the one in the diagram.

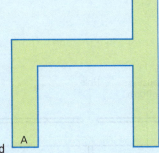

She introduced a woodlouse into the maze at A. The woodlouse is forced to make a right turn at the first junction. After this, the woodlouse has a choice of turning left or right. The student recorded the direction of the second turn for each woodlouse. The results are shown in the table.

Direction of turn	Number of woodlice
Left	27
Right	73

Questions

1 The student cleaned the maze with a clean cotton bud between each trial. Suggest why.

2 a Name the statistical test that you could carry out to check whether these results are significantly different.

   b Give a suitable null hypothesis.

3 Suggest the advantage to a woodlouse of turn alternation behaviour.

Simple reflexes

Figure 22 shows a **reflex arc**. Notice that a stimulus is detected by a receptor. Impulses pass along a sensory neurone to a relay neurone in the spinal cord. This sends impulses along a motor neurone to an effector, usually a muscle. For example, if a person touches a hot object, this may lead to a rapid withdrawal of the hand from the hot object. This does not involve conscious thought, as the brain is not involved. This simple reflex has the advantage of being rapid, protecting the person from immediate danger.

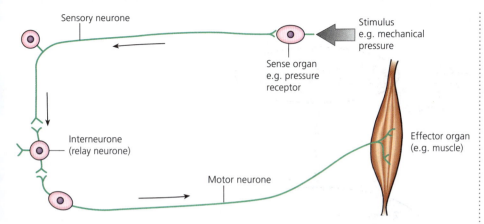

Figure 22 The layout of a reflex arc

Remember the features of a spinal reflex by the four 'i' words:

- innate (you are born with the reflex 'hard-wired' into the nervous system)
- invariable (you cannot modify it)
- immediate (it is very quick)
- involuntary (it does not require conscious thought)

Knowledge check 21

When the sensory neurone passes the impulse to the relay neurone in the spinal cord, it also sends impulses along a neurone that goes to the brain. However, this impulse does not reach the brain until the hand has already moved away from the hot object. What is the advantage of sending impulses to the brain?

Receptors

Receptors respond to various stimuli such as light, heat, touch or chemicals, but each type of receptor usually responds to just one specific type of stimulus.

Pacinian corpuscles are found in the skin and respond to pressure. They look like layers of membrane wrapped around the end of a sensory neurone, like the layers in an onion. You can see this in Figure 23.

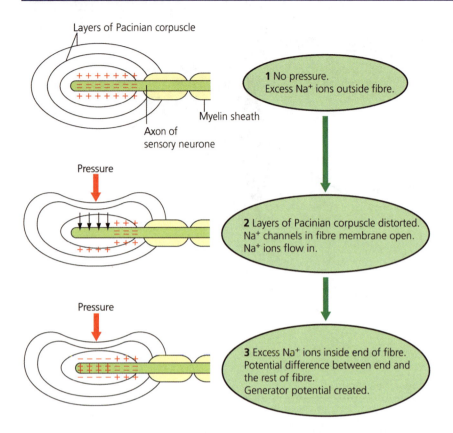

Figure 23 How a generator potential is set up in a Pacinian corpuscle

When the receptor is not being stimulated, there are many more sodium ions outside the axon than inside it. However, there are sodium ion channels in the membranes of the Pacinian corpuscle that open when they are stretched. They are therefore called **stretch-mediated channels**. When pressure is applied to the Pacinian corpuscle, the stretch-mediated sodium ion channels open and sodium ions enter the axon very quickly. This creates a **generator potential** and this causes nerve impulses to pass along the sensory neurone.

There are other receptor cells in the retina of the eye. The retina is a layer at the back of the eye. It contains two kinds of sensory cells:

- **Rod cells** contain the pigment rhodopsin. They are very sensitive and detect light at low intensities. These are more active in dim light.
- **Cone cells** contain the pigment iodopsin. These work best in bright light, but can detect colour. We have three kinds of cone, each sensitive to one wavelength, red, blue or green. The brain perceives colour as a result of the proportion of the different kinds of cone that are stimulated.

Rods and cones differ in their visual acuity. Figure 24 shows how the rods and cones are connected to the sensory neurones.

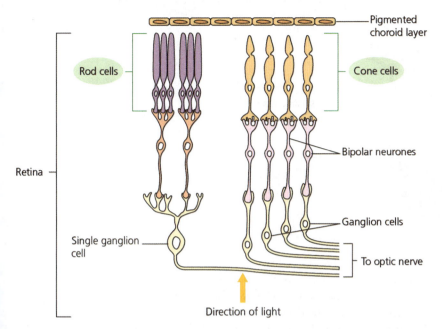

Figure 24 How the rods and cones are connected to other cells in the retina

You will see that the rods and cones connect to ganglion cells, which in turn connect to the optic nerve. However, cone cells are connected individually to ganglion cells and then to the optic nerve. This means that when a cone cell is stimulated by light, the brain can interpret exactly where the stimulus came from. Many rods share a connection to a single ganglion cell. There may be as many as 100 rod cells in one 'cluster'. This means that if one or two rods in a cluster are stimulated by light, the brain perceives the stimulus but cannot interpret exactly where it came from. This is the reason why you cannot read a newspaper in dim light.

Rod cells detect light intensity only, not colour. This is why, in dim light at night, you can see in black and white only. Cone cells detect colour. For example, if all three types of cone are stimulated equally, we perceive white. If red and green cones are stimulated equally, we perceive yellow.

Knowledge check 22

Nocturnal animals, which are active at night, have very few cone cells in the retina. Explain why.

Exam tip

The way that multiple rods are connected to a single ganglion cell is an example of spatial summation. This makes them more sensitive to light. See summation on page 36.

Control of heart rate

You will remember the structure of the heart from your first year of study. You will remember that the heart muscle is **myogenic**. This means that it contracts on its own, without any stimulation from the nervous system. Figure 25 shows the path of electrical activity in the heart during one heartbeat.

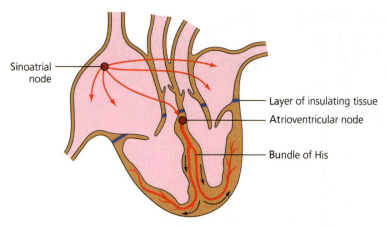

Figure 25 The route of electrical activity in the heart during one cardiac cycle

Impulses start at the sinoatrial node (SAN) in the wall of the right atrium. From here, impulses spread across the atria, causing atrial systole. Impulses cannot pass directly to the ventricles because there is a ring of fibrous tissue present. The impulses reach the atrioventricular node (AVN) where there is a short time delay, allowing the atria to empty completely. From here, the impulses pass down the bundle of His and along the Purkyne fibres, causing ventricular systole. After this, there is a short period of diastole when the heart muscle relaxes and there is no more electrical activity before the SAN stimulates another cardiac cycle.

However, our heart rate can change in response to the body's internal environment. When we exercise, the rate of respiration increases and more carbon dioxide is produced. Carbon dioxide dissolves in the blood to form carbonic acid, which is mildly acidic. This fall in pH is detected by **chemoreceptors** in the aorta, the carotid artery that takes oxygenated blood to the brain and in the medulla of the brain itself. These send more impulses to the cardiovascular centre in the medulla, which sends impulses along the sympathetic nerve to the SAN, increasing the rate at which the heart beats. When the pH rises again, this is detected by the chemoreceptors, which send impulses to the medulla. This time, more impulses are sent along the parasympathetic nerve to the SAN, slowing down the heart rate. This is summarised in Figure 26.

The heart rate also changes in response to blood pressure. When the body is exercising actively, body muscles contract strongly. This increases the rate at which deoxygenated blood is returned to the heart. This increased volume of blood stretches the heart muscle, causing the heart rate to increase and also causing the heart muscle to contract more strongly. This means that the heart pumps out a greater volume of blood each time it contracts. In other words, the stroke volume increases.

The increased stroke volume stretches the walls of the aorta and carotid arteries. This stimulates pressure receptors in the walls of these vessels, which send impulses to the cardiovascular centre in the medulla. As a result, the cardiovascular centre sends more impulses along the parasympathetic nerve to the SAN, slowing the heart rate.

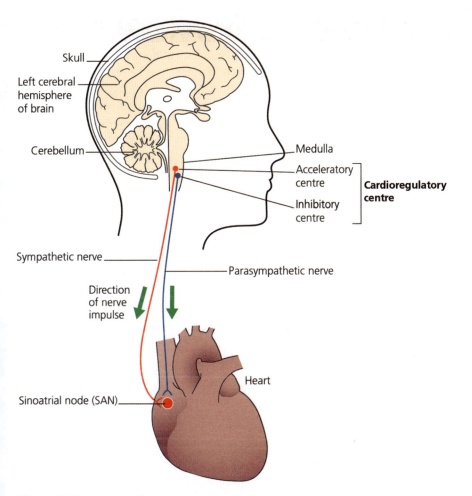

Skull

Left cerebral hemisphere of brain

Cerebellum

Medulla

Acceleratory centre

Inhibitory centre

Cardioregulatory centre

Sympathetic nerve

Parasympathetic nerve

Direction of nerve impulse

Heart

Sinoatrial node (SAN)

Figure 26 The control of heart rate

Knowledge check 25

Cardiac output (CO) is the volume of blood pumped from the heart in a minute. Stroke volume (V) is the volume of blood pumped from the heart in one cardiac cycle. Trained athletes can increase their stroke volume to a much higher figure than a non-athlete. In an investigation of an Olympic medal-winning cross-country skier, it was found that this athlete's cardiac output was $38 \, dm^3$ and the stroke volume was $210 \, cm^3$ during peak performance. Use the formula:

$CO = R \times V$ (where R = heart rate)

to find the athlete's heart rate.

Knowledge check 23

Explain how increasing heart rate reduces the carbon dioxide concentration in the blood.

Knowledge check 24

What effect does an increased carbon dioxide concentration have on haemoglobin? (You learned this in the first year of your course.)

Knowledge check 26

A trained athlete has a lower resting heart rate than a non-athlete. Suggest why.

Summary

- Organisms increase their chances of survival by responding to changes in their environment. In flowering plants, shoots and roots can grow towards or away from specific stimuli such as light, gravity and chemicals. These growth movements are called tropisms and result from the uneven distribution of indoleacetic acid (IAA).
- Mobile organisms show taxes and kineses. Taxes are directional movements in response to a directional stimulus. For example, the organism may move towards, away from or at an angle to a stimulus. A kinesis is when an organism moves faster and makes more turns when in less favourable conditions, while the rate of moving and turning slows when conditions are more favourable.

- A reflex arc involves a stimulus being detected by a receptor. Impulses pass along a sensory neurone to the spinal cord, where it synapses with a relay neurone. The relay neurone synapses with a motor neurone that stimulates an effector. This does not involve conscious thought.
- Receptors detect specific stimuli. Pacinian corpuscles in the skin respond to pressure. Rods and cones are found in the retina and respond to light. Heart muscle is myogenic, but the rate at which the heart beats can be modified by the nervous system. Chemoreceptors and baroreceptors detect changes in carbon dioxide concentration and blood pressure, sending impulses to the cardiovascular centre in the medulla to change the heart rate.

Nervous coordination

Nerve impulses

Figure 27 shows the structure of a motor neurone.

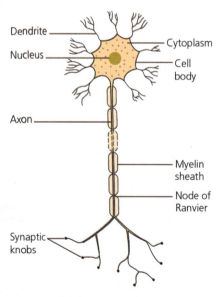

Figure 27 The structure of a motor neurone

This carries impulses from the central nervous system (brain and spinal cord) to effectors such as glands and muscles. Impulses enter the cell via the dendrites. The cell body contains most of the cytoplasm and organelles, and there is an axon, which can be as long as a metre in length. Many motor neurones, like the one in Figure 27, have a myelin sheath around them. The myelin sheath is made of Schwann cells, which wrap around the axon, providing many layers of fatty membrane that insulate the axon from surrounding cells and tissue fluid. Impulses pass down the axon to the synaptic knobs, where the impulse passes to the muscle at a **neuromuscular junction**.

Impulses are waves of electrical activity that pass along a neurone. When a neurone is at rest, its membrane is **polarised**. This means that there is a **potential difference** across the membrane, i.e. the inside of the membrane is negative compared with the outside. This is achieved by the distribution of ions. In the membrane of the axon are carrier proteins called **sodium–potassium pumps**. These actively transport three sodium ions out of the axon in return for two potassium ions moving into the axon. Since both ions are positively charged, this means that there are more positive ions outside the axon than inside. In addition, more potassium ions 'leak' out of the axon than sodium ions leak in. This results in a potential difference of about −70 mV inside the axon compared with outside. This is called the **resting potential**.

There are other proteins in the membrane surrounding the axon. These are **sodium channels** and **potassium channels**. You can see how these work in Figure 28. These are voltage gated channels that allow ions to pass through them by facilitated diffusion when the potential difference across the membrane changes and they are open.

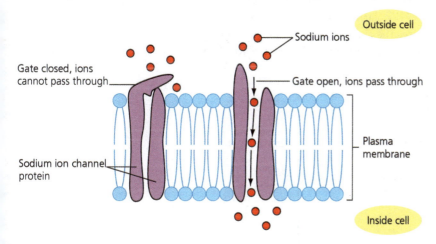

Figure 28 Gated sodium ion channels in the plasma membrane

Stimulation of a neurone causes a small change in the potential difference across the axon membrane. This causes the sodium channels to open. Sodium ions rush into the axon, down their electrochemical gradient. (This means they move in down the electrical gradient, as it is negative inside the axon, and down their concentration gradient.) So many sodium ions enter that the axon becomes positive compared with the outside. This stage is called **depolarisation**. At this point, the sodium channels close and the potassium channels open. These are also voltage-gated. Potassium ions rush out of the axon, down their electrochemical gradient. Slightly more potassium ions move out than sodium ions that entered, so the membrane briefly becomes hyperpolarised. This stage is called **repolarisation**. The sodium–potassium pump redistributes the ions and the membrane quickly returns to its resting potential. There is a short period after the sodium and potassium channels have closed when they cannot re-open. This is called the **refractory period**. This whole process is called an **action potential**, shown in Figure 29.

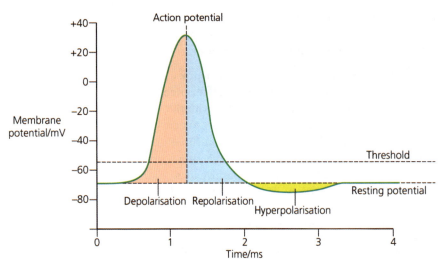

Figure 29 An action potential

Impulses occur when action potentials pass down a neurone. The depolarisation of one small section of a neurone sets off depolarisation in the next section of the neurone because the depolarisation causes the sodium gated channels to open. A wave of depolarisation passes along a neurone, rather like a Mexican wave. This is shown in Figure 30.

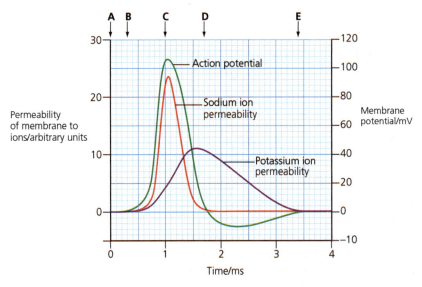

Figure 30 Changes in membrane potential during an action potential

As long as a stimulus is large enough to reach a **threshold value** it will lead to an action potential. A large stimulus does not cause a bigger action potential than a small one. This is called the **all-or-nothing principle**. Either a stimulus causes an action potential, or it doesn't. The strength of a stimulus is conveyed by the **frequency** of action potentials.

Action potentials pass along a specific axon at a constant speed. They do not speed up or slow down. However:

- wide diameter fibres conduct impulses faster than narrow fibres
- myelinated fibres conduct fibres faster than non-myelinated fibres

Non-myelinated axons conduct impulses in the way described above. Myelinated axons can exchange ions only in the gaps between the myelin, called nodes of Ranvier. This makes ion exchange more efficient so conduction is faster. You can see this in Figure 31.

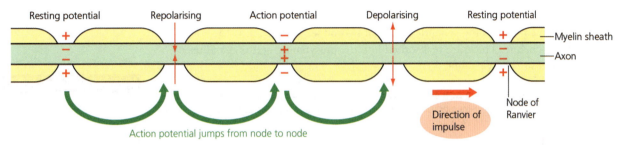

Figure 31 Saltatory conduction in a myelinated neurone

Conduction of nerve impulses is slower in cold conditions and faster when it is warmer. This is not an issue in mammals that regulate their body temperature but does have an effect in ectotherms. Temperature affects the kinetic energy of the sodium and potassium ions undergoing facilitated diffusion.

Synaptic transmission

When two neurones meet, they do not actually touch — there is a tiny gap between them called a **synapse**. You can see the structure of a synapse in Figure 32.

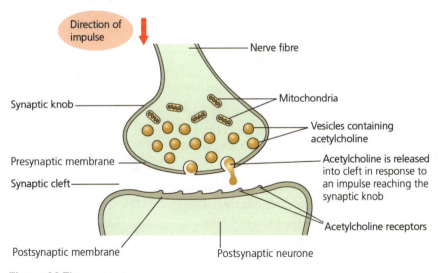

Figure 32 The synapse

Figure 33 shows a **cholinergic** synapse, which means it secretes acetylcholine as its transmitter substance.

Knowledge check 27

The refractory period ensures that impulses can pass along a neurone in one direction only. Explain why.

Knowledge check 28

Give two differences between active transport and facilitated diffusion.

- When an action potential arrives in the presynaptic neurone's synaptic bulb, calcium ion channels open and calcium ions enter the neurone.
- This causes vesicles of acetylcholine to migrate towards the presynaptic membrane and some of the vesicles fuse with the membrane.
- Acetylcholine is released into the synaptic cleft. It diffuses across the synaptic cleft.
- Acetylcholine fits into specific receptor proteins on the postsynaptic membrane, opening sodium ion channels.
- Sodium ions enter the postsynaptic neurone. If sufficient sodium ions enter to reach the threshold value, this sets up an action potential in the postsynaptic neurone.
- There is an enzyme in the synaptic cleft that hydrolyses acetylcholine to acetyl and choline. These are reabsorbed into the presynaptic neurone and used to re-synthesise acetylcholine.

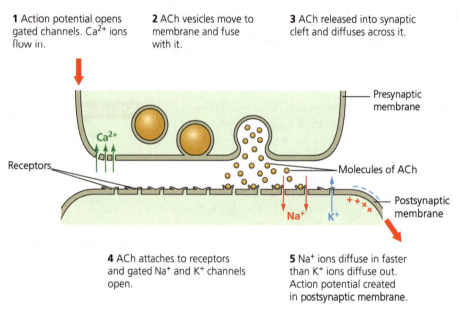

1 Action potential opens gated channels. Ca^{2+} ions flow in.

2 ACh vesicles move to membrane and fuse with it.

3 ACh released into synaptic cleft and diffuses across it.

4 ACh attaches to receptors and gated Na^+ and K^+ channels open.

5 Na^+ ions diffuse in faster than K^+ ions diffuse out. Action potential created in postsynaptic membrane.

Figure 33 Sequence of events during an impulse transmission at a cholinergic synapse

Often, when one presynaptic neurone stimulates a postsynaptic neurone, too few sodium ions enter the postsynaptic membrane to overcome the threshold value, so the postsynaptic neurone does not 'fire'. However, sometimes action potentials can 'add up' to stimulate the postsynaptic neurone. This is called **summation**.

Spatial summation (see Exam tip on page 29) occurs when several presynaptic neurones receive an action potential at the same time, so that between them enough sodium ions enter the postsynaptic membrane to overcome the threshold value.

Temporal summation occurs when one presynaptic neurone receives action potentials in very quick succession, so that between them enough sodium ions enter the postsynaptic neurone to overcome the threshold value.

Synapses can also be inhibitory. At these synapses, an action potential in a presynaptic neurone makes it less likely that an action potential will be stimulated

Knowledge check 29

Synapses allow impulses to pass in one direction only. Explain how.

Knowledge check 30

Suggest two reasons why there are many mitochondria in the synaptic knob of the presynaptic neurone.

Knowledge check 31

What would happen if the enzyme was not present in the synaptic cleft to break down acetylcholine?

Exam tip

The rod cells in the retina are an example of spatial summation. This allows the rods to be sensitive to low light intensity.

in the postsynaptic neurone. One kind of inhibitory transmitter substance is GABA. GABA opens channels on the postsynaptic membrane that allow negatively charged chloride ions into the postsynaptic neurone. This makes the inside of the neurone even more negative compared with outside and therefore an action potential is much less likely to occur.

Many drugs operate at synapses. There are several ways in which they work, such as:
- some fit into the receptor proteins on the postsynaptic membrane and allow sodium ions in — these stimulate the synapse, e.g. nicotine in tobacco
- some fit into receptor proteins on the postsynaptic membrane but do not open the sodium channels — these act as a synaptic block, e.g. some kinds of local anaesthetic
- some inhibit the enzyme in the synaptic cleft that breaks down the transmitter substance

When a motor neurone synapses with a muscle fibre, there is a special kind of synapse called a **neuromuscular junction**. This is shown in Figure 34.

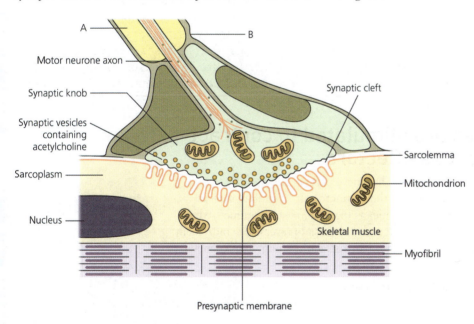

Figure 34 The neuromuscular junction

It works in a similar way to a normal synapse and there are only a couple of small differences:
- Neuromuscular junctions are always excitatory.
- Neuromuscular junctions always use acetylcholine, but synapses can use various transmitter substances.

When an action potential arrives at the synaptic bulb of the motor neurone:
- calcium ions enter the neurone and cause vesicles of acetylcholine to move towards the membrane
- acetylcholine is released into the synaptic cleft

- acetylcholine diffuses across the synaptic cleft and fits into specific receptor proteins on the postsynaptic membrane (the sarcolemma of the muscle fibre)
- sodium channels open on the sarcolemma and sodium ions diffuse in
- this depolarises the sarcolemma and allows calcium ions to enter the muscle fibre from the sarcoplasmic reticulum, which leads to muscle contraction

Summary

- When a neurone is not conducting an impulse, it is said to be at rest.
- It has a potential difference across the membrane caused by more positively charged ions being present outside the cell than inside. This is achieved by a sodium–potassium pump in the membrane.
- An action potential is started when the membrane becomes depolarised. Sodium channels open and sodium ions enter the neurone. This is followed by potassium channels opening and potassium ions leaving the neurone. This is repolarisation.
- Action potentials pass along neurones at a constant speed, although wide neurones and myelinated neurones are faster. There is a refractory period following the action potential, which separates each action potential.
- The all-or-nothing principle states that all action potentials are the same size; stronger stimuli are conveyed by a higher frequency of action potentials.
- Synapses are tiny gaps where neurones communicate. Transmitter substances cross the synaptic cleft and may stimulate an action potential in the postsynaptic neurone.
- Many drugs affect synapses. A neuromuscular junction is where a motor neurone stimulates a muscle. This works in a similar way to a synapse.

Skeletal muscles are stimulated to contract by nerves and act as effectors

Muscles are attached to bones by tendons. They move parts of the body by contracting — this means they can pull but they cannot push. Therefore skeletal muscles come in **antagonistic pairs**. This means that one muscle contracts while the other relaxes to bring about a movement, and vice versa to reverse the movement. The biceps and triceps muscles in the arm form an antagonistic pair. This is shown in Figure 35.

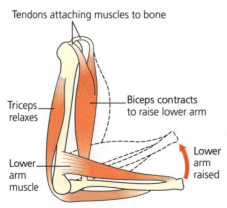

Tendons attaching muscles to bone

Triceps relaxes

Biceps contracts to raise lower arm

Lower arm muscle

Lower arm raised

Figure 35 An antagonistic pair of muscles

Skeletal muscles are made of muscle fibres. This is shown in Figure 36. Each muscle fibre is made of many myofibrils, surrounded by cytoplasm (called **sarcoplasm** in muscle tissue). The membrane around the muscle fibre is called the **sarcolemma**.

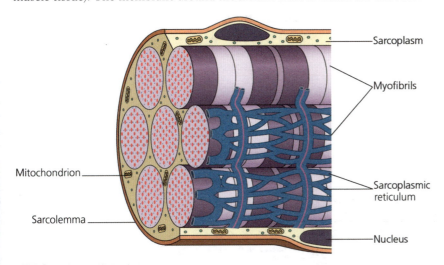

Figure 36 Structure of a muscle fibre

The fibre also contains a network of tubules called the **sarcoplasmic reticulum**, which stores calcium ions. The myofibrils appear banded when you look at them under the microscope. The bands are units called **sarcomeres**, containing thin filaments of **actin** and thick filaments made of **myosin**. They are arranged as shown in Figure 37.

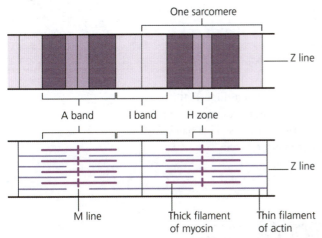

Figure 37 The structure of a sarcomere

You will see that the darkest part of the sarcomere is where both actin and myosin are present, and the lightest part is where there is only actin. Figure 38 shows what the sarcomere would look like if they were cut through in different places.

a)   b)   c)

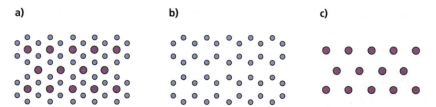

Figure 38 Sections through the sarcomere in different places showing the arrangement of the actin and myosin filaments

The myosin filaments form cross-bridges with the actin filaments. When the muscle contracts, the actin and myosin filaments slide over each other so that they overlap more. This can be seen in Figure 39.

Relaxed

A band   I band   H zone

Z line

Sarcomere   M line
Myosin   Myosin   Actin
head   filament filament

Contracted

Figure 39 Relaxed and contracted sarcomeres

The myosin molecule has lots of 'heads' sticking out. You can see this in Figure 40. The actin filaments have binding sites where myosin heads can bind. However, there is also a molecule called tropomyosin twisted around the actin filaments. This blocks the actin binding sites. This is what enables the muscle to remain relaxed.

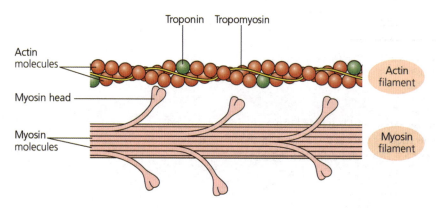

Troponin   Tropomyosin

Actin
molecules

Actin
filament

Myosin head

Myosin
molecules

Myosin
filament

Figure 40 Actin and myosin filaments

Knowledge check 33

When a sarcomere contracts, what happens to:

a the width of the I band?

b the width of the A band?

c the width of the H zone?

Figure 41 shows what happens when the muscle contracts.

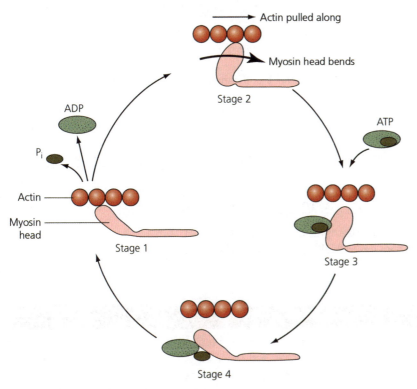

Figure 41 Stages in the cycle of actinomyosin cross-bridge formation

- When an action potential arrives at the neuromuscular junction, the sarcolemma is depolarised and sodium ions enter.
- This causes calcium ions to diffuse out of the sarcoplasmic reticulum into the myofibrils.
- Calcium ions attach to tropomyosin and cause it to change shape, making it twist away from the myosin-binding sites on the actin filaments.
- The myosin heads attach to the binding sites on actin, forming actinomyosin bridges.
- As this cross-bridge forms, the myosin head releases ADP and P_i and bends, pulling the actin molecule along for a short distance (this is called the power stroke).
- An ATP molecule attaches to each myosin head, separating it from the actin and making it change shape.
- The ATP is hydrolysed into ADP and P_i and the myosin head is straightened (this is called the recovery stroke).
- The myosin heads repeat this process by attaching to another actin-binding site and sliding the actin along a little more.

There are different types of muscle fibres. Slow fibres are able to keep working for long periods. They respire aerobically and do not fatigue easily. Fast fibres have fewer capillaries and mitochondria, and tend to respire anaerobically. Table 1 shows the key difference.

Knowledge check 34

Name the type of reaction in which ATP is broken down into ADP and P_i.

Table 1 Different types of muscle fibres

Slow muscle fibres	Fast muscle fibres
Long contraction–relaxation cycle	Short contraction–relaxation cycle
Smaller store of calcium ions in sarcoplasmic reticulum	Large store of calcium ions and more sarcoplasmic reticulum
Dense network of blood capillaries around fibres for supply of oxygen and glucose for aerobic respiration	Fewer blood capillaries around fibres
ATP largely obtained from aerobic respiration	ATP largely obtained from anaerobic respiration
Many, large mitochondria, nearer the surface of the fibres	Fewer, smaller mitochondria, more evenly distributed
Small amount of glycogen	Larger amount of glycogen and phosphocreatine
Slower rate of ATP hydrolysis in myosin heads	Higher rate of ATP hydrolysis in myosin heads, so more actinomyosin bridges formed per second
Resistant to fatigue, since less lactate is formed	Quickly become fatigued, since more lactate is formed

Muscle fibres also contain stores of phosphocreatine. This can be used to produce ATP very rapidly by transferring a phosphate ion to ADP, replacing the ATP that has been hydrolysed.

ADP + phosphocreatine → ATP + creatine

Summary

- Skeletal muscles work in antagonistic pairs.
- Muscles are made up of muscle fibres, which are made of bundles of myofibrils. The myofibrils are composed of sarcomeres containing actin and myosin.
- When muscle contracts, actinomyosin cross-bridges break, then re-form as one filament slides past the other. ATP provides the energy used in cross-bridge formation.
- Tropomyosin blocks the actin-binding sites in relaxed muscles. For muscle contraction to take place, calcium ions bind to tropomyosin, which causes the binding sites to be exposed.
- There are two kinds of muscle fibre: fast twitch and slow twitch.

Homeostasis is the maintenance of a stable internal environment

Principles of homeostasis and negative feedback

Homeostasis is the maintenance of a stable internal environment. Mammals maintain a stable body temperature and pH because this creates an optimum environment for enzyme activity. Remember that enzymes are proteins. If the temperature is too low, there will be less kinetic energy and fewer enzyme–substrate complexes formed. If the temperature is too high, or the pH too high or too low, weak hydrogen and ionic bonds holding the enzyme in its tertiary structure will break. This means the enzyme's active site will change so that the substrate does not fit so easily. If the temperature increases too much, or the pH alters too much, the enzymes will be denatured.

Blood glucose concentration is kept relatively stable to ensure that cells have an adequate supply of glucose for respiration. However, if blood glucose concentration is

Knowledge check 35

Explain why disulfide bridges are less likely to break than hydrogen or ionic bonds if the temperature increases (or the pH changes).

Knowledge check 36

Controlling heart rate is one way of controlling blood pH. Explain how.

too high, this lowers the water potential of the blood. Water moves into the blood by osmosis, increasing blood pressure. This can cause damage to blood vessels leading to such problems as heart disease and blindness.

Body temperature, pH and blood glucose concentration are all maintained by **negative feedback**. The principle of negative feedback is shown in Figure 42.

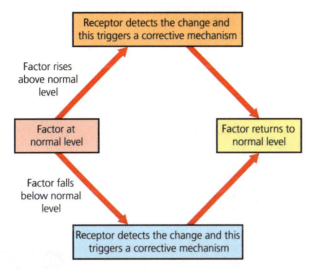

Figure 42 The principle of negative feedback

In **positive feedback**, the stimulus brings about a response that changes the factor even further away from the normal level.

Control of blood glucose concentration

When blood glucose concentration rises above the normal range, for example following a meal, this is detected by **beta cells (β cells)** in the islets of Langerhans in the pancreas. Insulin reduces blood glucose concentration by:

■ attaching to receptor proteins in the cell membranes of liver, muscle and adipose tissue, stimulating an increased number of channel proteins to be present in those cells to take in glucose by facilitated diffusion

■ stimulating enzymes that convert glucose to glycogen — this process is called **glycogenesis**

■ activating enzymes in adipose tissue that convert glucose to fatty acids and glycerol, that can be stored as fat

When blood glucose concentration falls, this is detected by **alpha cells (α cells)** in the islets of Langerhans. These cells respond by secreting another hormone, **glucagon**. Glucagon increases blood glucose concentration by:

■ attaching to receptor proteins in cell membranes and activating enzymes in the liver and muscle that convert glycogen back to glucose — this process is called **glycogenolysis**

■ activating enzymes that convert other substances such as lactate and amino acids to glucose — this is called **gluconeogenesis**

The whole process is summarised in Figure 43.

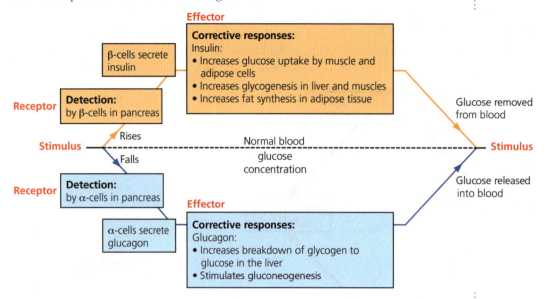

Figure 43 The control of blood glucose concentration

Exam tip

Be careful that you can spell these words correctly — glucagon, glycogen, gluconeogenesis, glycogenolysis and glycogenesis. Examiners will not allow mis-spellings of these words because they are so similar yet mean very different things.

Another hormone that increases blood glucose concentration is adrenaline. This hormone is released at times of stress in the 'fight or flight' response. Adrenaline attaches to receptor proteins on the surface of liver and muscle cells, activating enzymes that convert glycogen to glucose.

Adrenaline and glucagon do not have a direct effect on liver cells like insulin does. Instead, they bind to a receptor protein in the cell membrane and cause a cascade of reactions within the cell. This is called the **second messenger model**, shown in Figure 44.

Adrenaline and glucagon work in a similar way. They bind to a specific receptor protein in the cell membrane, which activates an enzyme in the cell membrane called adenyl cyclase. This removes two phosphate groups from ATP, producing cyclic adenosine monophosphate (cAMP). This activates another enzyme, protein kinase, which converts glycogen to glucose phosphate. Note that this means that one molecule of hormone can lead to many molecules of cAMP being formed and in turn many more enzyme molecules.

Knowledge check 37

Why is it an advantage to have two hormones to control blood glucose concentration?

Knowledge check 38

What are the features of glycogen that make it a good storage molecule? (You learned this in year 1 of your course.)

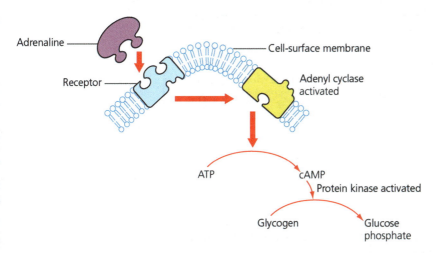

Figure 44 The second messenger model of hormone action

Diabetes is a condition in which people are unable to control their blood glucose concentration. It usually develops in childhood or early adulthood. In type 1 diabetes, people no longer produce insulin. This means that they may have excessively high blood glucose concentrations (hyperglycaemia) or extremely low blood glucose concentrations (hypoglycaemia). In both these states it is possible that a diabetic may go into a coma and even die. However, the condition can be controlled by eating a balanced diet, exercising and taking insulin injections after meals.

Type 2 diabetes is more likely to develop in older people, and in people who are obese and eat a diet high in sugar. People with type 2 diabetes usually produce insulin as normal but their cells have stopped responding to it. In its early stages it can be treated by losing weight, exercising regularly and eating a balanced diet that is low in sugars, high in fibre and in which the carbohydrates are mainly complex polysaccharides that are digested slowly.

> **Knowledge check 39**
>
> What advice about diet and exercise do you think would be given to a person with type 2 diabetes?

Exam tip

You may be asked to evaluate the advice given to people about healthy eating, and the role of the food industry in relation to the increased incidence of type 2 diabetes. For example, you may be asked to evaluate food labelling that informs people how much sugar is present and whether food manufacturers should be forced to reduce the amount of sugar and fat in processed foods.

Required practical 11

Production of a dilution series of a glucose solution

A student made up a dilution series of a glucose solution. She was given a 5% glucose solution and used it to make up 20 cm^3 of 0%, 1%, 2%, 3%, 4% and 5% glucose solutions. She also had 20 cm^3 of a glucose solution with an unknown glucose concentration. She added 1 cm^3 of quantitative Benedict's reagent to each tube and heated them in a boiling water bath for exactly 1 minute.

➡

After this time, she tested the glucose solutions in a colorimeter. A colorimeter measures the colour of a solution by measuring the percentage of light of a specific wavelength shone on the sample that passes through. The darker the colour, the less light is transmitted by the solution and the more light is absorbed. The diagram shows how a colorimeter works.

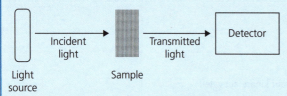

The student put a small sample of each solution in a cuvette and measured the percentage of light that the solution absorbed. She used the results to produce a calibration curve. This is shown below.

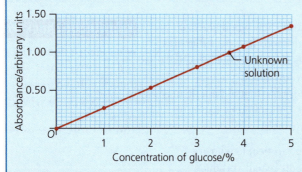

Then the student placed a sample of the unknown solution in a cuvette and used the calibration curve to find its glucose concentration.

Questions

1 Describe how you would make up a 5% glucose solution.

2 Complete the table to show how you would make up $20\,cm^3$ of each glucose solution, using a 5% glucose solution and distilled water.

Concentration of solution/%	5% glucose solution/cm³	Distilled water/cm³
5	20	0
4		
3		
2		
1		
0	0	20

3 Explain why the same volume of Benedict's reagent was added to each tube and the tubes were heated for the same length of time.

4 Describe how the student could use the calibration curve to find the concentration of glucose in the unknown sample.

Control of blood water potential

Osmoregulation is the way in which blood water potential is regulated. This involves the kidney. The kidneys contain thousands of nephrons. You can see the structure of a nephron in Figure 45.

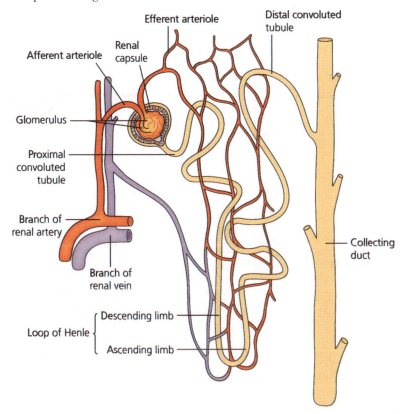

Figure 45 Structure of a nephron

The **glomerulus** carries out ultrafiltration. Blood enters the glomerulus under pressure, since it has come from the renal artery. The afferent vessel is narrower than the efferent vessel, increasing the pressure. Water from the blood plasma, along with smaller molecules and ions, is forced out through the wall of the capillary, the basement membrane and the glomerulus wall. Blood cells and larger molecules remain in the blood but the filtrate enters the nephron.

At the proximal convoluted tubule, glucose is absorbed into the blood vessels surrounding the nephron by active transport unless the glucose concentration is excessively high (as can happen in the case of a person with diabetes). Amino acids are all absorbed by active transport, and water is reabsorbed by osmosis until the filtrate is isotonic with the blood. Some mineral ions and urea are also absorbed.

The loop of Henle creates a gradient of sodium ions deep in the medulla of the kidney. Sodium chloride diffuses out of the lower part of the ascending limb. Chloride ions are actively transported out of the thick ascending limb, while sodium ions follow passively. This creates a low water potential in the medulla of the kidney.

The filtrate entering the descending limb is already isotonic with the blood, but as it passes down the descending limb it encounters an increasingly low water potential in the medulla of the kidney. Therefore more water leaves the descending limb by osmosis into the medulla and the blood vessels surrounding the nephron. The filtrate reaches its lowest water potential at the base of the loop of Henle. The ascending limb of the loop of Henle is impermeable to water, so it remains reduced in volume and concentrated in salts. The loop of Henle is shown in Figure 46.

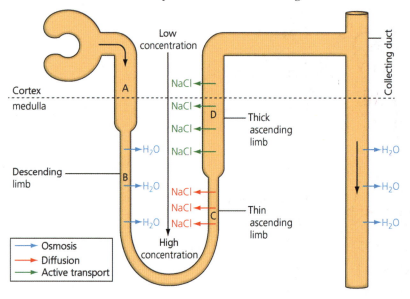

Figure 46 The loop of Henle

Knowledge check 40

The loop of Henle is an example of a countercurrent exchange system. You met this idea in year 1 when you learned about gas exchange in fish gills. Explain why this system in the loop of Henle is described as a countercurrent exchange system and how this is an advantage.

Most of the useful substances, such as glucose and amino acids, have already been reabsorbed from the filtrate. At the distal convoluted tubule, some of the remaining salts and water are reabsorbed. The hormone ADH (antidiuretic hormone) controls the reabsorption of water by affecting the permeability of the distal tubule and collecting duct to water. This is shown in Figure 47.

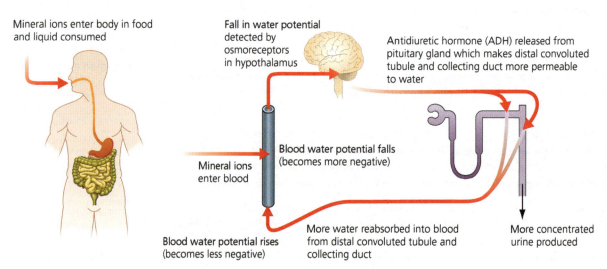

Figure 47 The action of antidiuretic hormone

Osmoreceptors in the hypothalamus of the brain are sensitive to the water potential of the blood flowing through it. When the water potential of the blood falls, the osmoreceptors send impulses to the pituitary gland, which secretes ADH. ADH increases the permeability of the distal convoluted tubule and the collecting duct by binding to protein channels called **aquaporins** in the cell surface membranes of the cells lining these parts of the nephron. This allows more water to leave the filtrate by osmosis. Therefore a more concentrated urine is produced.

If the water potential of the blood is high, the osmoreceptors send impulses to the pituitary gland, inhibiting the production of ADH. This causes the aquaporins to leave the cell-surface membrane of cells lining the distal convoluted tubule and collecting duct. Therefore less water is reabsorbed and the urine is more dilute. This is another example of negative feedback. This is summarised in Figure 48.

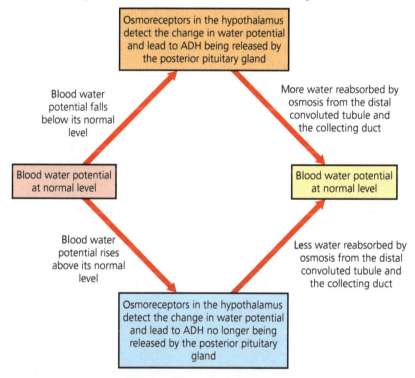

Figure 48 The control of blood water potential by the kidney

Knowledge check 41

During labour, a hormone called oxytocin is released from the pituitary gland and intensifies and speeds up contractions. The increase in contractions causes more oxytocin to be released and the cycle goes on until the baby is born.

Explain how this is an example of positive feedback.

Content Guidance

Summary

- Blood glucose concentration is detected by cells in the islets of Langerhans.
- If blood glucose concentration is too high, insulin is released, which causes glucose to be stored as glycogen in liver and muscle cells and as body fat.
- If blood glucose concentration is too low, glucagon is released, which causes glycogen to be converted to glucose and stimulates the conversion of other molecules, especially glycerol and amino acids, to glucose. This is an example of negative feedback.
- Adrenaline is another hormone that increases blood glucose concentration in the 'fight or flight' response. Both adrenaline and glucagon work by the second messenger model.
- In diabetes, people are unable to control their blood glucose concentration.
- Type 1 diabetes is caused by a failure of the pancreas to produce enough insulin.
- Type 2 diabetes occurs when enough insulin is present but the body fails to respond, sometimes because there are not enough insulin receptors.
- Nephrons in the kidney carry out ultrafiltration of the blood. All the useful molecules in the blood, such as glucose and amino acids, are reabsorbed from the filtrate under normal conditions.
- The loop of Henle provides a high concentration of sodium ions in the medulla of the kidney, resulting in more water being reabsorbed from the filtrate by osmosis.
- The hypothalamus of the brain contains osmoreceptors, which monitor the water potential of the blood.
- When the water potential of the blood is too low, the osmoreceptors stimulate the release of antidiuretic hormone (ADH) from the pituitary gland, which increases the permeability of the distal convoluted tubule and collecting duct to water. Therefore more water is reabsorbed from the filtrate by osmosis, resulting in more concentrated urine.
- When the water potential of the blood is high, the osmoreceptors detect this and send impulses to the pituitary gland to inhibit the release of ADH. This results in more dilute urine being produced.

Questions & Answers

Exam format

For A-level biology, your exams will be structured as follows:

Paper 1	Paper 2	Paper 3
Any content from topics 1–4, including relevant practical skills	Any content from topics 5–8, including relevant practical skills	Any content from topics 1–8, including relevant practical skills
Written exam, 2 hours 91 marks, worth 35% of A-level	Written exam, 2 hours 91 marks, worth 35% of A-level	Written exam, 2 hours 78 marks, worth 30% of A-level
76 marks: mixture of long- and short-answer questions 15 marks: extended response	76 marks: mixture of long- and short-answer questions 15 marks: comprehension	38 marks: structured questions, including practical techniques 15 marks: critical analysis of experimental data 25 marks: essay from a choice of two titles

The topics in this book are examined in paper 2 (together with topics 7 and 8) and in paper 3 (together with topics 1–4 and 7–8). Essay questions typical of sample paper 3 are not included in this guide but are included in the fourth student guide of this series as they require knowledge of topics 1–8.

Tips for answering questions

Use the mark allocation. Generally, 1 mark is allocated for one fact, concept or item in an explanation. Make sure your answer reflects the number of marks available.

Respond appropriately to the command words in each question, i.e. the verb the examiner uses. The terms most commonly used are explained below:

- **Describe** — this means 'tell me about…' or, sometimes, 'turn the pattern shown in the diagram/graph/table into words'; you should not give an explanation.
- **Explain** — give biological reasons for *why* or *how* something is happening.
- **Calculate** — add, subtract, multiply, divide (do some kind of sum!) and show how you got your answer — *always* show your working.
- **Compare** — give similarities *and* differences between…
- **Complete** — add to a diagram, graph, flowchart or table.
- **Name** — give the name of a structure/molecule/organism etc.
- **Suggest** — give a plausible biological explanation for something; this term is often used when testing understanding of concepts in an unfamiliar context.
- **Use** — you must find and include in your answer relevant information from the passage/diagram/graph/table or other form of data.

About this section

There are several ways of using this section. You could:

- 'hide' the answers to each question and try the question yourself. It needn't be a memory test — use your notes to see if you can make all the points you ought to make
- check your answers against the students' responses and make an estimate of the likely standard of your response to each question
- check your answers against the comments to see where you might have failed to gain marks
- check your answers against the terms used in the question — for example, did you *explain* when you were asked to or did you merely *describe*?

Comments

Student responses are followed by detailed comments. These are preceded by the icon **e** and indicate where credit is due. In the weaker answers, they also point out areas for improvement, specific problems and common errors, such as lack of clarity, weak or non-existent development, irrelevance, misinterpretation of the question and mistaken meanings of terms.

■ Paper 2-type questions

This paper is in the style of paper 2, but the real paper 2 will contain questions from all areas of the A-level course studied in year 2. Because this paper is in a revision guide for topics 5 and 6, these questions focus specifically on topics 5 and 6 with some parts of questions from year 1 of the course.

Question 1

Figure 1 represents energy flow through an ecosystem.

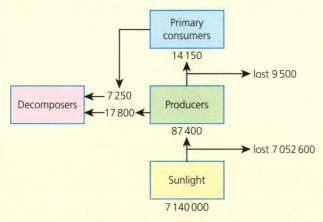

Figure 1

(a) Give suitable units for energy flow in Figure 1. (AO1) (2 marks)

> **Student A**
>
> **(a)** $kJ\,m^{-2}\,y^{-1}$

ⓔ This gets both marks — 1 mark for the energy unit and area unit, and 1 mark for a time unit. Any energy, area or time unit would be acceptable.

> **Student B**
>
> **(a)** $J\,ha^{-1}$

ⓔ This gets just 1 mark because there is no time unit.

(b) Give two ways in which energy is lost between producers and primary consumers. (AO1) (2 marks)

> **Student A**
>
> **(b)** In respiration, and not all the plant can be eaten such as the roots.

ⓔ This gets both marks for two sensible answers.

> **Student B**
>
> **(b)** Lost as heat and in faeces.

ⓔ This also gets both marks for two sensible answers.

(c) Calculate the percentage of available sunlight trapped by the producers. Show your working. (AO2)   (1 mark)

> **Student A**
>
> **(c)** Energy in sunlight = 7 140 000
>
> Energy in producers = 87 400
>
> % of energy trapped = 87 400/7 140 000 × 100%
>
> = 1.22%

ⓔ This gets both marks for a correct answer.

> **Student B**
>
> **(c)** 12%

ⓔ This gets no marks. It looks as though there is an arithmetic error — if student B had shown their working, the correct method might have scored 1 mark.

(d) Give two reasons why very little of the sunlight that falls on the producers is transferred into chemical energy in biomass in the producers. (AO1)   (2 marks)

> **Student A**
>
> **(d)** Some of the light is the wrong wavelength and some of it lands on a part of the plant that contains no chloroplasts.

ⓔ This gets both marks for two correct responses.

> **Student B**
>
> **(d)** Some is reflected and some lands on the bare earth.

ⓔ This gets only 1 mark, for the first point. The second point cannot be credited because the question asks about light that lands on the producer, so it excludes light landing on the bare earth.

Question 2

(a) Some liver was homogenised in ice-cold isotonic buffer solution. The mitochondria were separated by centrifugation. Explain why:

  (i) the solution was isotonic. (AO1)   (2 marks)

  (ii) a buffer solution was used. (AO1)   (2 marks)

Note that this question contains some biology from the first year of your course in a question about respiration, which is from the second year of the course.

Student A

(a) (i) It has the same water potential as the cell so the mitochondria won't take in water by osmosis and burst.

  (ii) This keeps pH constant so the enzymes in the mitochondria will not be denatured as a result of pH changes.

ⓔ This gets all 4 marks — 2 marks for each section — for a completely correct response.

Student B

(a) (i) It has the same water potential as the cell so the cell won't burst by osmosis.

  (ii) To keep pH constant.

ⓔ This gets 1 mark for (i) by explaining what isotonic means, but not the second mark. We don't need to keep the cell intact as we're homogenising it — it's the mitochondrion that needs to be intact. Part (ii) also gains 1 mark — for explaining what the buffer does — but the second mark hasn't been awarded as the link to preventing denaturation of enzymes isn't made.

(b) The mitochondria were used to investigate part of the respiration pathway. A respiratory substrate and inorganic phosphate were added to the mitochondria. A short time later, ADP was added to the mixture. An oxygen probe was used to record the oxygen concentration. Figure 2 shows the results.

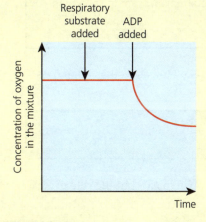

Figure 2

(i) **Name a suitable respiratory substrate for this investigation. Give a reason for your answer. (AO2)** (2 marks)

Student A

(b) (i) Pyruvate because it can be broken down in the mitochondrion in the link reaction.

🅮 This gets 2 marks, one for a correct substrate and one for a correct reason. Note that the student could have given the name of any intermediate that comes after pyruvate as well, such as acetyl coA.

Student B

(b) (i) Glucose because that is the usual respiratory substrate in the cell.

🅮 This gets no marks. Glucose is wrong because this is broken down in the cytoplasm, and the enzymes that do this will not be present in a suspension of mitochondria.

(b) (ii) **Inorganic phosphate was added to the mixture. Explain why. (AO1)** (1 mark)

Student A

(b) (ii) This is needed to join on to ADP to make ATP.

🅮 This gets the mark for a clear answer.

Student B

(b) (ii) It is needed for ATP synthesis.

🅮 This is also correct and gains the mark.

(b) (iii) **Explain the change in the concentration of oxygen when the ADP was added. (AO2)** (2 marks)

Student A

(b) (iii) Oxygen is used as the final acceptor in the electron transfer chain, in which ATP is made. So oxygen and phosphate are used up when ADP is converted to ATP.

🅮 This gets both marks as the role of the oxygen is stated as well as linking its use to ADP being added.

Student B

(b) (iii) Oxygen is needed to convert ADP to ATP.

🅮 This gets 1 mark for linking to the formation of ADP to ATP but the role of oxygen is not clear enough for the other mark.

Question 3

Some students carried out an investigation into gas exchange in an aquatic plant. Four tubes were set up, each containing 20 cm^3 of an indicator solution. This indicator solution is:

- yellow below pH 6
- green between pH 6.1 and pH 7.5
- blue at pH 7.6 and above

Similar sized pieces of an aquatic plant were placed into three of the tubes and bungs were added. They were set up as shown in Figure 3 and left 10 cm from a bench lamp for 1 hour.

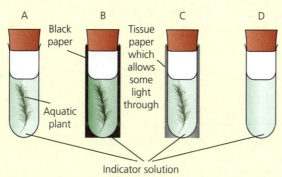

Figure 3

Table 1 shows the colour of the indicator in each tube at the beginning and end of the investigation.

Table 1

Tube	Conditions	Colour of indicator at start	Colour of indicator at end
A	Uncovered	Green	Blue
B	Surrounded by black paper	Green	Yellow
C	Covered with tissue paper	Green	Green
D	Uncovered	Green	Green

(a) Explain why:

    (i) a bung was placed in each tube. (AO3)   (1 mark)

    (ii) the tubes were all placed the same distance from the lamp. (AO3)   (1 mark)

> **Student A**
>
> **(a) (i)** To prevent carbon dioxide from the air changing the colour of the indicator.
>
>   **(ii)** So that the heat effect of the light was the same for all the tubes.

ⓔ These answers gain both marks.

Student B

(a) (i) To prevent evaporation of the liquid.

(ii) So that the light intensity was the same for all the tubes.

ⓔ These answers do not score any marks. Usually bungs prevent evaporation, though this is not really going to make much difference over 1 hour. The important reason here is to prevent carbon dioxide in the air changing the colour of the indicator and to ensure that carbon dioxide produced in the tube remains in the tube and changes the colour of the indicator. Also, the tubes do not receive equal light intensity since one is covered in black card and another in tissue paper. However, if the tubes were at different distances from the light they would be exposed to different temperatures.

(b) Explain the results for:
tube A
tube B
tube C (AO2)

(6 marks)

Student A

(b) Tube A. Photosynthesis is occurring faster than respiration, so carbon dioxide is taken in overall and oxygen released. This makes the indicator more alkaline.

Tube B. This plant can't photosynthesise as there is no light so it is respiring only. This produces carbon dioxide which makes the pH lower.

Tube C. The low light intensity means that respiration and photosynthesis are occurring at similar rates. Overall there is no net production or take-up of carbon dioxide so the pH stays the same.

ⓔ These answers score full marks. The student recognises that the plant respires all the time, but in A photosynthesis is greater than respiration. Also, the student recognises that removing carbon dioxide which is acidic makes the indicator show a more alkaline pH. Similarly in B, the student explains why respiration alone is happening and relates the pH change to the production of carbon dioxide. In C, the student can explain the net gas exchange in terms of both respiration and photosynthesis.

Student B

(b) Tube A. The plant is photosynthesising so it takes in carbon dioxide. This makes the pH rise.

Tube B. The plant is respiring because it is in the dark so it releases carbon dioxide which makes the indicator more acid.

Tube C. This plant is photosynthesising slowly so it is not taking in much carbon dioxide so it does not change the pH.

ⓔ These answers are incomplete so overall only 2 marks are awarded. In A, the student has not recognised that respiration is also happening and that there is *net* production of carbon dioxide. However, 1 mark can be awarded for the idea that carbon dioxide is acidic and changes the colour of the indicator. In B, the candidate gets the first mark for respiring — not carrying out photosynthesis is implied in the 'because it is in the dark' — but the second mark has already been awarded once. In C there is no mention of respiration so the first mark can't be awarded and the second mark can't be awarded because there is no recognition that there is no *net* change in carbon dioxide because respiration and photosynthesis balance each other.

(c) Explain the purpose of tube D. (AO3) (2 marks)

Student A

(c) This is a control to compare with the other tubes. It shows that the indicator doesn't change colour just because it is exposed to light.

ⓔ This gets 2 marks for the idea of comparison and for explaining that it is there to eliminate the possibility that the indicator changes colour because it is exposed to light.

Student B

(c) This is a control and shows it is the plant that makes a difference.

ⓔ This gets no marks as it doesn't mention comparison. Second, although it says it shows the plant is making a difference, this is not linked to gas exchange.

Question 4

(a) Figure 4 shows a relaxed sarcomere.

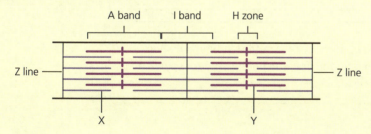

Figure 4

(i) Name:

X

Y (AO1) (2 marks)

Student A

(a) (i) X actin

Y myosin

ⓔ This scores both marks.

Student B

(a) (i) X is actin

Y is myosin

🄴 This also scores both marks.

(a) (ii) Give two pieces of evidence from Figure 4 that show that the sarcomere is relaxed. (AO1) (2 marks)

Student A

(a) (ii) The I band is quite wide and also the H zone so the actin and myosin are not overlapped as much as they could be.

🄴 This is a good answer scoring both marks and even adding some extra detail about the overlap of the actin and myosin.

Student B

(a) (ii) I band and H zone are wide.

🄴 These two features score a mark each.

(b) Describe the role of the following in muscle contraction:

(i) calcium ions (AO1) (2 marks)

(ii) ATP (AO1) (2 marks)

Student A

(b) (i) They bind to tropomyosin and move it, revealing the actin binding sites.

(ii) This provides the energy needed to form actin–myosin cross-bridges.

🄴 These answers get full marks. In (i) there is a mark for binding to tropomyosin and also for exposing the binding sites on the actin. In (ii) the student says that ATP provides energy and says what this energy is needed for.

Student B

(b) (i) They expose the actin binding sites by binding to tropomyosin.

(ii) ATP provides the energy to break and re-form the actin–myosin cross-bridges.

🄴 These answers get full marks too in the same way as student A. However, these are expressed a little differently.

Question 5

(a) What is a tropism? (AO1)   (1 mark)

> **Student A**
>
> **(a)** It is a directional growth response to a directional stimulus.

e This gets full marks.

> **Student B**
>
> **(a)** It is a movement in response to a directional stimulus.

e This doesn't get any marks. Although a directional stimulus is mentioned, it says 'movement' rather than growth. This definition describes a taxis, not a tropism.

(b) In an investigation, a scientist grew two coleoptiles. He removed a segment from each coleoptile and placed them in darkness, as shown in Figure 5(a). Figure 5(b) shows the appearance of the coleoptiles after 2 days.

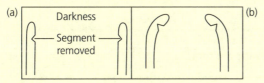

Figure 5

Explain these results. (AO2)   (4 marks)

> **Student A**
>
> **(b)** IAA is produced at the top of the coleoptile and diffuses downwards. As the segment has been removed, IAA can't travel down that side but travels normally down the other side. The side with more IAA (where the segment hasn't been taken away) has more IAA, so the cells elongate more that side.

e This gets full marks. The student has explained clearly that auxin/IAA is produced at the tip and explains its unequal distribution. This is linked to cell elongation and growth.

> **Student B**
>
> **(b)** There are more cells on one side of the coleoptile so they grow more. The coleoptiles bend towards the light.

e This does not get any marks. The student has not understood what is happening. There is no reference to IAA and how it is unequally distributed. The student has not understood that the coleoptiles are in darkness, so suggests they are growing towards light.

Question 6

(a) **Figure 6 shows part of the nitrogen cycle.**

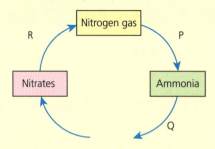

Figure 6

Name processes P, Q and R. (AO1) (3 marks)

Student A

(a) P = nitrogen fixation

Q = nitrification

R = denitrification

ⓔ This answer gets all 3 marks for being fully correct.

Student B

(a) P = ammonification

Q = nitrification

R = denitrification

ⓔ This answer gets 2 marks, but the first response is wrong. The student saw the term 'ammonia' and linked this to ammonification — but in ammonification the ammonia derives from proteins and other organic material, not from nitrogen gas.

(b) **Name two nitrogen-containing compounds in living organisms. (AO1)** (1 mark)

Student A

(b) Proteins and DNA

ⓔ This gets 1 mark as both suggestions are correct.

Student B

(b) Amino acids and starch

ⓔ Although amino acids is correct, starch is wrong, so this gets no marks.

(c) (i) Explain why dairy farmers need to add fertiliser to their fields to maintain production of milk. (AO2)

(2 marks)

Student A

(c) (i) They need to make the grass grow as fast as possible to feed the cattle, and as the milk produced is removed, not all the nutrients are recycled.

ⓔ This gets 2 marks, for explaining that the fertiliser is used to make sure the grass grows well, in addition to explaining how nutrients overall are lost from the system when the crop or product is removed.

Student B

(c) (i) Nutrients are lost when the milk is sold.

ⓔ This gets 1 mark for explaining the loss of nutrients but how the fertiliser links to milk production is not explained.

(c) (ii) Give two advantages of using natural fertiliser rather than artificial fertiliser. (AO1)

(2 marks)

Student A

(c) (ii) It improves the quality of the soil and is less likely to leach out.

ⓔ This gets both marks for two clear advantages.

Student B

(c) (ii) It is cheaper and breaks down gradually.

ⓔ This gets 1 mark for breaking down slowly, but cheaper on its own gets no marks. There needs to be a valid reason, such as 'it is cheaper because the farmer needs to get rid of the manure from the dairy cattle on the farm anyway'.

Question 7

(a) People with a condition called diabetes insipidus do not produce sufficient antidiuretic hormone. They produce large amounts of dilute urine. Explain why. (AO2)

(3 marks)

Student A

(a) ADH causes more water to be removed from the distal convoluted tubule and collecting duct by opening aquaporins in the membrane. If there is no ADH then more water is lost in the urine.

ⓔ This gets both marks for a full and detailed answer.

Questions & Answers

Student B

(a) If there is no ADH then less water is reabsorbed in the distal convoluted tubule and collecting duct.

e This gets 1 mark for saying the area where ADH operates but there is no reference to how it affects the permeability so no second mark.

(b) People with diabetes mellitus may have a very high glucose concentration in their blood. This results in glucose being present in their urine. Explain why. (AO2)　　　　　　　　　　　　　　　　　　　　　　(3 marks)

Student A

(b) Glucose is filtered out of the blood at the glomerulus. It is reabsorbed by active transport in the proximal convoluted tubule. If the blood glucose concentration is too high, there are not enough carrier proteins to reabsorb all the glucose so some of it is left in the filtrate and excreted in the urine.

e This gets full marks for a fully detailed answer.

Student B

(b) If there is too much glucose, it can't all get reabsorbed in the proximal tubule so some stays in the filtrate and ends up in the urine.

e This gets 1 mark for the idea that the glucose can't all be reabsorbed so it stays in the filtrate, and the proximal tubule is mentioned, which gains a second mark. However, there is no reference to carrier proteins or active transport.

(c) Figure 7 shows the factors that control the release of the hormone oxytocin. Oxytocin stimulates uterine contractions during childbirth.

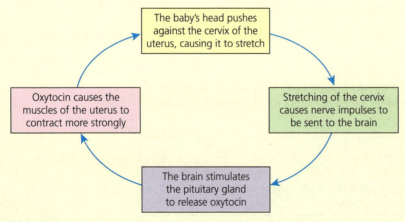

Figure 7

(i) Use Figure 7 to explain what is meant by an effector. (AO2)　　　　(1 mark)

(ii) This is an example of positive feedback. Use Figure 7 to explain why. (AO2)　　(2 marks)

Student A

(c) (i) Pituitary gland because this releases oxytocin when stimulated by the brain.

(ii) The stimulus (stretching of the cervix) brings about a response that causes the muscles of the uterus to contract even more.

e This gains both marks for a clear and correct answer.

Student B

(c) (i) Muscles of the uterus as they produce a response.

(ii) The stimulus causes more oxytocin to be released, which increases uterine contractions.

e This also gains both marks. Note that there is more than one effector in the example given.

Question 8

(a) The three-toed sloth lives in trees, hanging upside-down from branches. They sleep for almost 10 hours a day and move very slowly. Their muscles contain a high proportion of slow-twitch fibres. Explain the advantage of this to the sloth. (AO2)

(3 marks)

Student A

(a) Slow-twitch fibres respire aerobically so they do not fatigue easily and work for a long period of time. They have a lot of mitochondria and capillaries to supply the fibres with oxygen and glucose, so aerobic respiration can occur which supplies oxygen more efficiently.

e This is a full answer as the advantage is given but the explanation is also given so it gains full marks.

Student B

(a) They can stay contracted for a long time without fatiguing.

e This answer gives the advantage to the sloth but it is not explained, so there is only 1 mark.

(b) Three-toed sloths belong to the family Bradipodidae. Name three other taxonomic groups that they will share. (AO1)

(2 marks)

Student A

(b) Kingdom, phylum, class

e This gains 2 marks for three correct answers.

> **Student B**
>
> **(b)** Class, order, genus

e This gains just 1 mark, as class and order are correct but genus isn't.

(c) Sloths feed on leaves which are composed mainly of cellulose. Sloths have stomachs with four compartments containing many bacteria. The food stays in the stomach much longer than it does in most other herbivores such as cattle. Suggest the advantage to the sloth of:

  (i) having bacteria in the stomach (AO2)   (2 marks)

  (ii) food staying in the stomach for a long time (AO2)   (1 mark)

> **Student A**
>
> **(c) (i)** The bacteria secrete enzymes that digest the cellulose. This is useful to the sloth as higher organisms do not produce a cellulase enzyme.
>
> **(ii)** This allows more time for the cellulose and other substances to be digested, and for the products of digestion to be absorbed, so overall the sloth gains more nutrients.

e (c) (i) Both marks here for realising they must digest cellulose and knowing that most animals do not have a cellulase enzyme.

   (ii) This gets the mark as the student understands that more time for digestion means more nutrients can be absorbed.

> **Student B**
>
> **(c) (i)** They digest the cellulose which the sloth probably can't digest as it probably doesn't have a cellulase enzyme.
>
> **(ii)** This allows more time for the bacteria to digest the plant material and for the nutrients to be absorbed.

e (c) (i) Both marks are awarded here for understanding that the sloth doesn't have cellulase and the bacteria digest the cellulose into sugars that the sloth can use.

   (ii) Again this gets the mark for realising that digestion takes time and it allows more time for absorption.

(d) Sloths have long shaggy fur that holds water. Algae (single-celled photosynthetic organisms) grow in this fur. Suggest the advantage of this to:

  (i) the algae (AO2)   (1 mark)

  (ii) the sloth (AO2)   (1 mark)

Student A

(d) (i) The algae have a moist surface to grow on that is exposed to light so the algae can photosynthesise.

(ii) The sloth is camouflaged as it is similar in colour to the leaves of the trees.

ⓔ These answers gain both marks for being clear and correct.

Student B

(c) (i) This is a surface for the algae to grow on and there is more light for photosynthesis high up in the trees where the sloth lives.

(ii) The algae provide nutrients to the sloth.

ⓔ The first answer gains a mark but the algae don't provide nutrients to the sloth. This might be arguable if the student had suggested the sloth grooms itself and eats the algae from the fur, but the answer as it stands does not suggest a valid advantage to the sloth.

Question 9

Beta blockers are drugs that fit into receptors on the sinoatrial node of the heart. They prevent the neurotransmitter noradrenaline from binding. They also prevent the hormone adrenaline from binding.

(a) Explain the effect beta blockers have on heart rate. (AO2)   (2 marks)

Student A

(a) They slow down the heart rate because the sympathetic nerve secretes noradrenaline, which stimulates the SAN to beat faster. Also adrenaline increases heart rate. Beta blockers stop them stimulating heart rate so it stays slow.

ⓔ This is a detailed answer and gains both marks.

Student B

(a) They stop adrenaline and the sympathetic neurone from speeding up the heart rate.

ⓔ This gets both marks but only just. The student has said that beta blockers slow the heart rate, or at least stop it speeding up, and links this to the action of adrenaline and the sympathetic neurone.

Doctors carried out an investigation to find out whether beta blockers improve the condition of people with congestive heart failure (CHF). They recruited a large number of men and women aged 40–80 and put them into two groups. One group was given regular doses of a beta blocker and the control group was given a placebo. The doctors examined the patients regularly over 18 months.

Some of the results are shown in Table 2.

Table 2

	Group given beta blocker	Control group
Number in group	399	396
Number of deaths from all causes	155	203
% alive after 18 months		

(b) (i) Complete the table by calculating the percentage of each group still alive after 18 months. Show your working. (AO2)   (2 marks)

Student A

(b) (i)

	Group given beta blocker	Control group
Number in group	399	396
Number of deaths from all causes	155	203
% alive after 18 months	38.8	51.3

$\frac{155}{399} \times 100 = 38.8\%$

$\frac{203}{396} \times 100 = 51.3\%$

ⓔ This gets both marks for a fully correct response.

Student B

(b) (i)

	Group given beta blocker	Control group
Number in group	399	396
Number of deaths from all causes	155	203
% alive after 18 months	38.84	51.26

ⓔ This gets both marks but this student has not shown the working. If the answers had been wrong, the examiner could not award any marks for a correct method.

(b) (ii) The doctors calculated that the risk of dying was reduced by 39% in the group given a beta blocker. They carried out a statistical test which was significant at $p < 0.001$. Explain what this means. (AO3)   (2 marks)

Student A

(b) (ii) This means that the chances of the differences between the two groups being due to chance is less than 0.1% so the difference is statistically significant.

e This gets both marks, for explaining that the differences are statistically significant and for explaining the *p* value.

Student B

(b) (ii) The results are not due to chance. There is a 0.001 chance of getting these results by chance.

e This gets no marks. The student refers to 'the results' rather than the differences between the groups. The student does not mention statistical significance and the *p* value shows a *less than* 0.001 probability of the differences being due to chance.

(c) (i) This investigation was double blind. Explain what this means and why it is important. (AO3) (2 marks)

(ii) Explain the purpose of the control group. (AO3) (2 marks)

Student A

(c) (i) Neither the patients nor the doctor know who is getting the beta blocker and who is getting the placebo. This is so that psychological factors cannot affect the results.

(ii) This is for comparison with the experimental group, so that they know how many deaths would occur when no beta blocker is given.

e This gets all 4 marks for a fully correct answer.

Student B

(c) (i) This is to stop bias from the doctors or the patients because neither of them knows who is getting the placebo or the drug.

(ii) This shows the difference the beta blocker makes and allows you to compare the results.

e This gets both marks in (i) but only one in (ii) because the student doesn't explain that we need to know how many deaths would occur without the beta blocker.

Question 10

Read the passage and use the information in the passage and your own knowledge to answer the questions.

Botox is a brand name for botulinum A toxin. This is one of a series of toxins produced by the bacterium *Clostridium botulinum* that each fit into different receptor proteins. Botulinum A toxin attaches itself to a protein called SNAP-25 in the presynaptic membrane at the neuromuscular junction and prevents acetylcholine being released.

Botox is widely used for cosmetic treatments since it can smooth out wrinkles. In a study to show its safety, scientists labelled it with a radioactive isotope of iodine. They found that the toxin did not spread far from the site of injection. Botulinum toxin is very powerful. Just one molecule can block a synapse for several days. In another study, scientists looked for fragments of damaged SNAP-25 after they had injected rats' whisker muscles with botulinum A. They found evidence of damaged receptors in the medulla of the rats' brains. However, clinicians who administer Botox have said that the toxin does not travel far from the site of injection when used correctly, and that this local spread can be an advantage.

Epilepsy is a nervous system disorder in which groups of neurones deep in the brain transmit impulses abnormally. One group of researchers has found that administering one botulinum injection into one group of neurones can stop the whole group of neurones from being active. Botulinum toxin can also be used to treat other disorders, such as blepharospasm, in which there are abnormal spasms of the eyelid muscles.

People who are infected with *Clostridium botulinum*, either from infected food or an infected wound, may suffer fatal paralysis. One way to treat this disease is to administer antibodies against botulinum toxin.

(a) **Botulinum A toxin attaches to the SNAP-25 protein but not any other receptors (lines 3–4). Explain why. (AO2)** (2 marks)

> **Student A**
>
> (a) The receptor protein SNAP-25 has a precise tertiary structure that only botulinum A toxin fits into.

🅔 This gets both marks for a fully correct answer.

> **Student B**
>
> (a) Botulinum A toxin is exactly the right shape to fit into the SNAP-25 protein but other receptors are a different shape so it won't fit into them.

e This gets both marks as well. It makes the same points as the answer above, just expressed a little differently.

(b) Explain how Botox causes muscle paralysis. (AO1)   (3 marks)

> **Student A**
>
> **(b)** It stops acetylcholine being released, so the postsynaptic membrane is not stimulated. Therefore sodium ions do not enter the muscle and calcium ions are not released so the muscle can't contract.

e This gets all 3 marks for a fully correct and detailed answer.

> **Student B**
>
> **(b)** If acetylcholine isn't released, sodium channels on the sarcolemma don't open. Therefore sodium ions don't enter the muscle and the muscle doesn't contract.

e This gets all 3 marks too, although it is a little less detailed than student A's answer.

(c) Labelling Botox with radioactive iodine allowed scientists to see whether it spreads from the site of injection (lines 8–9). Explain how. (AO3)   (1 mark)

> **Student A**
>
> **(c)** Radioactivity allows the Botox to be detected using photographic film, for example. Samples of tissues may be taken from different sites and their radioactivity assessed.

e This detailed answer gets the mark.

> **Student B**
>
> **(c)** The scientists will measure the level of radioactivity at different sites at different distances from the site of injection.

e This gets the mark, although it is less detailed than student A's answer.

(d) (i) If botulinum toxin damages synapses in the medulla, this could be fatal. Explain why. (AO2)   (1 mark)

   (ii) Give one advantage and one disadvantage of testing Botox on rats (lines 10–13). (AO3)   (2 marks)

> **Student A**
>
> **(d) (i)** The medulla controls vital activities, such as breathing or changing heart rate.
>
>   **(ii)** This means we know that it works and it's likely to be safe without testing on humans where it might cause harm. However, rats are different from humans so we can't be sure it's safe for humans.

Questions & Answers

e This gets all marks for both sections.

Student B

(d) (i) The medulla has a lot of important control centres. The heart wouldn't beat without it.

(ii) You can check it isn't seriously harmful without testing on humans but many people think it's unethical to test a drug on animals, especially when it's used for cosmetic reasons.

---

e This gets only 1 mark in (i) for the idea that the medulla has control centres in it. But it is untrue that the heart wouldn't beat without it, as the heart is myogenic. It also gets both marks in (ii).

(e) **Explain how antibodies might be useful to treat a person paralysed by botulinum toxin. (AO1)** (3 marks)

Student A

(e) Antibodies have specifically shaped receptor sites. An antibody that is the right shape to fit onto botulinum toxin would join with the toxin and stop it binding to the membrane receptors. Then the muscle could contract as usual.

---

e This gets full marks as it mentions the idea of shape and fit, explaining how the antibody prevents botulinum toxin causing paralysis.

Student B

(e) The antibodies would bind to the botulinum toxin, which stops it binding to the SNAP-25 receptor.

---

e This answer does not mention the idea of shape and fit or specificity. This gains only 1 mark.

(f) **A journalist who had read this information wrote an article entitled 'Botox should be banned'. Do you agree with this statement? Give reasons for your answer. (AO3)** (3 marks)

Student A

(f) Yes: because there is evidence it can spread from the site of injection and cause important muscles to be paralysed.

No: because it has been used a lot and has not caused harm in people. Also, there is evidence that it can be used to treat health disorders such as epilepsy.

ⓔ This gets full marks for three valid reasons that include both sides of the argument.

> **Student B**
>
> **(f)** Yes, because it is usually used to smooth out wrinkles, which is cosmetic, and there is some evidence it can spread to paralyse important muscles. However, this evidence is in rats and not in humans.

ⓔ This gets full marks too, again because there are three valid reasons and it includes both sides of the argument.

■ Paper 3-type questions

This paper is in the style of paper 3, but the real paper 3 will contain questions from all areas of the A-level course studied in year 2. Because this paper is in a revision guide for topics 5 and 6, these questions focus specifically on topics 5 and 6 with some parts of questions from year 1 of the course. Sample paper 3-type essay questions are not included in this guide but are included in the fourth student guide of this series as they require knowledge of topics 1–8.

Question 1

Some scientists carried out an investigation into the conditions that produce the best yield of cucumbers. They carried out the investigation in glasshouses. They varied light intensity, carbon dioxide concentration and temperature. Their results are shown in Table 1.

Table 1

Investigation	Light intensity/% of maximum	Temperature/°C	Carbon dioxide concentration/%	Yield/kg
A	15	20	0.03	14
B	15	20	0.13	19
C	50	20	0.03	24
D	50	20	0.13	49
E	100	20	0.03	25
F	100	20	0.13	59
G	15	30	0.03	14
H	15	30	0.13	21
I	50	30	0.03	24
J	50	30	0.13	59
K	100	30	0.03	25
L	100	30	0.13	79

(a) (i) Explain the importance of carrying out this investigation in glasshouses. (AO3)   (1 mark)

(ii) Give two other factors you would need to keep constant. (AO3)   (1 mark)

> **Student A**
>
> (a) (i) You can control all the variables such as temperature and carbon dioxide concentration in a glasshouse.
>
> (ii) Fertiliser and water availability.

ℯ This gets both marks.

> **Student B**
>
> **(a) (i)** If you grow cucumbers in a field you can't keep things like temperature and light intensity the same.
>
> **(ii)** Sowing density and soil nutrients.

ℯ This also gets both marks.

(b) Investigations A and G produced the same yield of cucumbers.
Explain why. (AO2)   (2 marks)

> **Student A**
>
> **(b)** In both cases carbon dioxide concentration was the limiting factor so although there was a higher temperature in G, this made no difference. B has a higher yield than G, but a lower temperature than G, so CO_2 is the limiting factor.

ℯ This gets both marks, for identifying the fact that carbon dioxide concentration was limiting, and giving a reason.

> **Student B**
>
> **(b)** Increasing the temperature to 30°C might have been too hot.

ℯ This gets no marks as the student has not looked closely enough at the data. The student has not realised that there is a limiting factor and hasn't identified it.

(c) A journalist wrote an article about this in a farming magazine. He said that farmers should grow cucumbers at 30°C, 0.13% carbon dioxide and maximum light intensity to produce a more profitable crop. Do you agree with this conclusion? Give reasons for your answer. (AO3)   (5 marks)

> **Student A**
>
> **(c)** Increasing light intensity makes the biggest difference, ✓ then carbon dioxide concentration and then temperature. ✓ F has 59 kg yield and L gives only 20 kg more for a 10°C rise in temperature. ✓ The farmer needs to consider the cost of increasing these factors. The cost of increasing the temperature 10°C may not be worth it unless the extra 20 kg of cucumbers fetch a high enough price. ✓ However, getting the greatest yield does not mean this crop is profitable, not only because of the cost of producing the cucumbers but also the size and shape of the cucumbers may affect the profit made. ✓

@ This gets full marks for a well-argued answer.

Student B

(c) The highest yield is under the conditions the journalist mentions so this is right. ✓ However, the farmer needs to think about the cost of producing these conditions. ✓ If he can increase the yield early or late in the year he will get a higher price and this will be more profitable. ✓

@ This gets 3 marks for several points but this student does not use data to back up the answer. However, this student does realise that the profitability of the crop depends not just on yield but also on when in the season the yield is produced.

Question 2

A technician set up a respirometer as shown in Figure 1. Potassium hydroxide solution absorbs carbon dioxide.

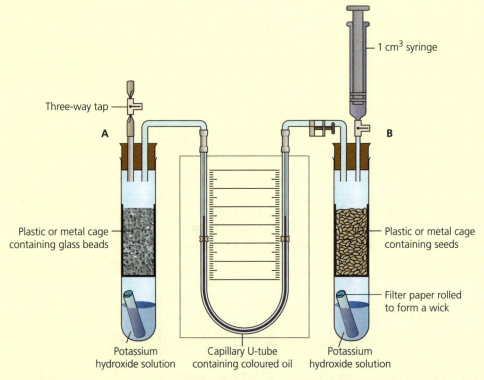

Figure 1

The tubes were placed in a thermostatically controlled water bath at 25°C for 5 minutes. The syringe was inserted and the syringe barrel moved until the fluid in the manometer was equal on both sides. The position of the barrel in the syringe and the fluid in the meniscus was noted. After 5 minutes, the barrel in the syringe was used to return the meniscus to the starting level. The new position of the syringe was noted. The mass of germinating seeds in tube B was recorded. This investigation was repeated at 10°C.

(a) (i) Why was potassium hydroxide present in both tubes? (AO3)   (1 mark)

Student A

(a) (i) So both tubes are the same except for the germinating seeds. ✔

e This gets full marks.

Student B

(a) (i) To absorb CO_2

e This gets no marks because this information was given in the question.

(a) (ii) Why was a filter paper wick placed in the potassium hydroxide
          solution? (AO2)   (1 mark)

Student A

(a) (ii) To increase the surface area for carbon dioxide absorption.

e This gets a mark for a correct answer.

Student B

(a) (ii) To absorb the carbon dioxide.

e This gets no marks because it does not explain the need for a wick.

(b) The apparatus was left in the water bath for 5 minutes before the syringe
     was inserted. Explain why. (AO3)   (1 mark)

Student A

(b) To allow the apparatus to equilibrate and all the contents had reached the
     temperature of the water bath.

e This gets full marks for a correct answer.

Student B

(b) To make sure the whole apparatus had reached the required temperature.

e This also gets full marks for a correct answer.

(c) Explain why the glass beads were necessary in tube A. (AO3)   (2 marks)

Student A

(c) This means the volume inside each tube is roughly the same so any change
     in temperature and pressure, that would affect the volume measured, ✔
     will be the same in both tubes. ✔

ℯ This gets full marks for a correct answer.

> **Student B**
>
> **(c)** As a control.

ℯ This is incorrect so no mark awarded.

(d) The seeds used 198 mm^3 of oxygen in 5 minutes at 25°C. The seeds weighed 28.3 g. Calculate the rate of oxygen consumption in mm^3 g^{-1} min^{-1}. Show your working. (AO2)   (2 marks)

> **Student A**
>
> **(d)** $\frac{198}{28.3}$ = 6.996 mm^3 g^{-1} in 5 minutes
>
>   = 1.399 mm^3 g^{-1} min^{-1} ✓✓

ℯ This gets full marks for a correct answer.

> **Student B**
>
> **(d)** 6.9

ℯ This is incorrect and appears to show an intermediate calculation. However, the student hasn't shown any working so cannot gain any credit for the method.

Question 3

A student cut three thin strips of muscle tissue (from fresh meat from a butcher). The strips were cut parallel to the muscle fibres. These thin strips were about 1 mm wide and about 40 mm long. Each strip was placed flat on a microscope slide and each was bathed in a solution of mineral ions. The mineral ions solution was then poured off. The length of each strip of muscle tissue was measured using a millimetre ruler and recorded.

A different solution was added to each slide and left for 5 minutes.
- Slide A had 0.5 cm^3 of ATP solution added.
- Slide B had 0.5 cm^3 of a boiled ATP solution added.
- Slide C had 0.5 cm^3 of an isotonic salt solution added.

After 5 minutes, the solution was removed, the muscle tissue strips were straightened and they were measured again. Table 2 records some of the results.

Table 2

Slide	Solution added	Original length/mm	Final length/mm	% change in length
A	ATP solution	38	32	
B	Boiled ATP solution	41	34	
C	Isotonic salt solution	39	38	

(a) Complete Table 2 by calculating the % change in length. (AO2) (2 marks)

Student A

(a)

Slide	Solution added	Original length/mm	Final length/mm	% change in length
A	ATP solution	38	32	15.8
B	Boiled ATP solution	41	34	17.1
C	Isotonic salt solution	39	38	2.6

ⓔ This gets full marks. Note that the student has used the correct formula to calculate this, i.e. (original length – final length)/original length × 100%.

Student B

(a)

Slide	Solution added	Original length/mm	Final length/mm	% change in length
A	ATP solution	38	32	84
B	Boiled ATP solution	41	34	83
C	Isotonic salt solution	39	38	97

ⓔ This gets no marks. This student has divided the final length by the original length and then multiplied by 100. This does not calculate the percentage change in length.

(b) (i) Explain the result for slide A. (AO2) (2 marks)

(ii) Slides A and B gave similar results. Explain why. (AO2) (2 marks)

Student A

(b) (i) ATP provided the energy ✓ to break and re-form the actin–myosin cross-bridges, causing muscle contraction. ✓

(ii) Boiling ATP makes no difference, so it still provides the energy for muscle contraction. ✓ ATP is not a protein so it is not denatured when boiled. ✓

ⓔ This gets full marks for a full and correct answer.

Student B

(b) (i) ATP provides the energy for muscle contraction. ✓

(ii) The muscle still contracted, showing ATP is not an enzyme and is not denatured ✓ on boiling so it still provides energy. ✓

e This gets only 1 mark in (i) because there is no reference to breaking or forming actin–myosin cross-bridges. However, in (ii) there is a full and correct answer getting both marks.

(c) Suggest the purpose of slide C. (AO3)   (2 marks)

> **Student A**
>
> **(c)** This is to compare with the others ✓ and shows the muscle does not contract on its own. ✓

e This gets both marks.

> **Student B**
>
> **(c)** As a control to compare with A and B. ✓

e This gets just 1 mark as it does not explain why we need a comparison.

Question 4

A group of students carried out an investigation to find the effect of different concentrations of IAA (auxin) on the growth of roots. They prepared conical flasks, each holding 50 cm^3 of a solution containing a different concentration of IAA.

(a) Given a solution of 10^{-4} M IAA and distilled water, describe how you would prepare solutions containing 10^{-5} M and 10^{-6} M IAA. (AO3)   (2 marks)

> **Student A**
>
> **(a)** I would pipette 10 cm^3 10^{-4} M IAA into a flask containing 90 cm^3 distilled water and mix them together. This is 10^{-5} M. Then I would pipette 10 cm^3 of the 10^{-5} M solution into 90 cm^3 distilled water to make a 10^{-6} M solution.

e This gets both marks as both solutions are correctly made and the volumes of auxin solution and distilled water are correct.

> **Student B**
>
> **(a)** I would take 1 cm^3 of the 10^{-4} M solution and add this to 10 cm^3 distilled water. Then I would take 1 cm^3 of this solution and add it to 10 cm^3 distilled water.

e This gets no marks, as the auxin solution would need to be added to 9 cm^3 distilled water rather than 10 cm^3. Notice also that the student has not said that the first solution would be 10^{-5} M and the second 10^{-6} M. This makes the answer more difficult for the examiner to evaluate.

The students placed 50 *Arabidopsis* seeds in each flask, placed a cotton wool bung in each and incubated them at 20°C under a uniform light source for 5 days. After this time, they measured the length of each root. The results are shown in Figure 2.

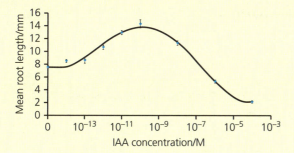

Figure 2 Mean root length of *Arabidopsis* seedlings grown in different IAA concentrations

(b) (i) What information do the standard deviation values give? (AO2)   (2 marks)

    (ii) Explain one possible advantage of growing the seeds in a solution in a flask rather than in soil. (AO3)   (2 marks)

Student A

(b) (i) This tells you how spread the measurements are around the mean and also tells you if the mean for one solution is significantly different from the mean for another solution.

    (ii) This means the only variable is the auxin concentration. If they had used soil there would be other nutrients present and then you wouldn't know whether the root length was caused by the auxin concentration or something in the soil.

---

e (b)

    (i) This gets the full 2 marks for a good explanation.

    (ii) This is also a valid advantage with a clear explanation, worth both marks.

Student B

(b) (i) This shows how spread about the mean the root lengths are.

    (ii) This makes it easier to remove the seedlings from the flask and measure their length. If they had to be pulled out of the soil, the root might break so the length you measure might be wrong.

---

e (b) (i) This gets 1 mark for understanding what standard deviation is, but the student has not mentioned that this can indicate whether the mean for one auxin concentration is significantly different from the mean for a different concentration.

    (ii) This is a different suggestion from the one that student A gave, but it is also valid. The student has clearly explained how damage to the root could occur if the seedlings were removed from soil, and this could result in inaccurate measurements.

(c) Describe the results of this investigation. (AO2) (3 marks)

> **Student A**
>
> **(c)** As auxin concentration increases to 10^{-15} M, root length increases. When auxin concentration increases above this, root growth is progressively inhibited, and at auxin concentrations above about 10^{-7} M, the root is shorter in length than when there is no added auxin in the solution.

🅔 This is a good answer worth all 3 marks, since the student has described the general pattern, has given a value for the peak of the graph and has noticed that some higher auxin concentrations produce less root growth than when the seedlings are grown in distilled water alone.

> **Student B**
>
> **(c)** As auxin concentration increases, so does root length until it reaches 14 mm, which is almost double the length of root when grown in distilled water alone. Above this concentration, root growth is inhibited.

🅔 This gets 2 marks — one for the general shape and one for a value from the graph where root length is a maximum. However, the student has not related the root length at higher auxin concentrations to the control with distilled water. Therefore a third mark cannot be awarded.

Required practicals answers

Required practical 7

1 So that it didn't dissolve in the solvent.

2 To break open the cell wall and membranes to release the pigments from the chloroplasts.

3 This produced a small, concentrated spot so that there was plenty of each pigment in the sample.

4 If the origin had been below the level of solvent, the pigments would have dissolved in the solvent instead of moving up the paper.

5 This stops the solvent evaporating and keeps the atmosphere in the tube saturated with solvent so that the spots can travel up the paper readily.

6 Distance moved by spot = 4 mm; distance moved by solvent front = 55 mm
Therefore $R_f = \dfrac{4}{55} = 0.073$

7 These pigments absorb light of a slightly different wavelength from that absorbed by chlorophyll and can pass on the light energy they absorb to chlorophyll.

Required practical 8

1 Chlorophyll in the chloroplasts absorbed light energy. This caused electrons to pass through the electron transfer chain and on to DCPIP. This caused the DCPIP to decolorise.

2 (a) This shows that DCPIP does not decolourise on exposure to light, but that the decolourisation occurs only when chloroplasts are present.
(b) This shows it is the chloroplasts, and not the isotonic saline they are suspended in, that causes the DCPIP to decolorise.

3 Isotonic saline has the same water potential as the stroma of the chloroplast. This means that there will be no net movement of water by osmosis into or out of the chloroplast. As a result, the chloroplast will not take in water and burst, nor will it lose water which might affect its metabolic activity.

4 The chloroplast suspension will contain some enzymes released from the organelles in the cell. These might digest the chloroplasts. Keeping the mixture ice-cold will stop these enzymes working. Also, keeping the mixture cold will ensure that the enzymes in the chloroplast remain stable until the investigation starts.

5 This would not affect the time taken for the DCPIP to decolourise, since this is an investigation of the light-dependent stage of photosynthesis only. Carbon dioxide is required for the light-independent stage.

6 The student could have taken samples from tube 1 at regular intervals and measured the transmission of light through the tube, using a colorimeter.

7 The student could set up several replicates of tube 1 and place them at different distances from the bench lamp. The student could record the time taken for each tube to decolourise.

Required practical 9

1 Carbon dioxide.

2 There would have been some respiratory substrates stored in the yeast cells.

3 Glucose can enter glycolysis directly. Sucrose needs to be digested into glucose and fructose first, which takes a little longer, therefore slightly less carbon dioxide is given off.

4 The yeast would have needed to synthesise enzymes to digest these substrates before being able to use them, which would take time. Therefore less carbon dioxide is given off.

5 This would be useful. It would show that the carbon dioxide given off comes from respiration (as boiled yeast would have denatured enzymes, therefore no carbon dioxide should be given off). It eliminates the possibility that carbon dioxide is given off by the respiratory substrate.

Required practical 10

1 To remove any chemical scent that might be left by the woodlouse, as otherwise the woodlice might be responding to chemicals and not direction.

2 (a) Chi-squared. (b) Following a forced right turn, a woodlouse is just as likely to turn right or left.

3 This would ensure the animal progresses in a roughly straight line and is always moving into new areas. If it continually turned the same way, it would be revisiting the same area it had already explored.

Required practical 11

1 Measure 5 g of glucose into a volumetric flask and make it up to $100\,cm^3$ using distilled water.

2

Concentration of solution/%	5% glucose solution/cm^3	Distilled water/cm^3
5	20	0
4	16	4
3	12	8
2	8	12
1	4	16
0	0	20

3 So that the colour of the solutions could be compared. This would make sure that difference in colour is the result of the different glucose concentrations and not any other factor.

4 The student would measure the % transmission of light through the unknown solution. They would find this value on the y-axis of the graph, then draw a horizontal line across the graph until it reached the line. Then they would draw a vertical line down to the x-axis. The point where the line meets the x-axis is the concentration of glucose in the unknown solution.

Knowledge check answers

1 a X, **b** Y

2 Reduced NADP and ATP are made in the light-dependent stage, and the light-independent stage cannot occur once supplies of these have run out.

3 GP gains hydrogen from reduced NADP when it is converted to TP, so therefore it is reduced.

4 a In graph C carbon dioxide concentration is the limiting factor, since graph B shows a higher rate of photosynthesis and the only difference is that the carbon dioxide concentration is higher. In graph B temperature is the limiting factor, since the rate of photosynthesis is higher in A and the only difference is that the temperature is higher.

b It is unlikely that temperature or carbon dioxide concentration are the limiting factor here as the rate of photosynthesis is already high. It is likely to be another factor, such as water availability or number of chloroplasts.

5 Pyruvate is oxidised to acetate, and NAD is reduced. Remember that oxidation is loss, reduction is gain. Pyruvate loses hydrogens and oxygen, while NAD gains hydrogen.

6 All the NAD in the cell would be reduced. Therefore there would be no NAD for the reaction in which triose phosphate is converted to pyruvate. If this reaction doesn't happen, no ATP can be produced and therefore the cell would die.

7 In anaerobic respiration, the glucose is only partly broken down (to pyruvate, a 3C compound). Also, in anaerobic respiration electron transfer does not happen. This is the stage when a great deal of ATP is made from the reduced coenzymes by oxidative phosphorylation.

8 It takes 4.2 J to raise the temperature of $1 cm^3$ water by 1°C. Therefore energy in willow =
$(4.2 \times 650 \times 3)/0.5 \,J\,g^{-1} = 16\,380 \,J\,g^{-1}$

9 a Any mass unit per area per time, e.g. $kg\,m^{-2}\,y^{-1}$

b and c Any energy unit per area per time, e.g. $kJ\,m^{-2}\,y^{-1}$

10 NPP of mature rain forest (Puerto Rico) is $54\,600 \,kJ\,m^{-2}\,y^{-1}$
GPP of lucerne (alfalfa) crop (USA) is $102\,480 \,kJ\,m^{-2}\,y^{-1}$

11 a $N = 3050 - 1900 - 1025$
$= 3050 - 2925 = 125$

b Energy per area per time, e.g. $kJ\,m^{-2}\,y^{-1}$

12 Cow cannot eat all the plant (e.g. roots); some primary production consumed by other animals, e.g. insects, rabbits.

13 There isn't enough energy left to support another trophic level.

14 A lot of the energy is lost in heat from respiration; some of the biomass eaten is indigestible.

15 DNA, RNA, ATP (any two). Allow correct intermediates in respiration and photosynthesis, e.g. GP, TP, RuBP or NADP.

16 The crop adds ammonium and nitrate ions to the soil, improving soil fertility; when the plant decomposes it provides humus/other nutrients, e.g. phosphorus; growing a green manure crop between other crops reduces weeds.

17 Nutrients are taken up from the soil by the crop and are lost when the crop is harvested. Therefore fresh nutrients need to be added for the new crop.

18 This increases the growth of the grass, so there is more food for the animals. The animals use the nutrients from the grass to increase their biomass. When the animals produce milk, or are slaughtered for meat, this removes nutrients, which need to be replaced.

19 a In A, auxin is evenly distributed in the stem because gravity is constantly changing. Therefore the cells elongate evenly. In B, there is more auxin on the lower side of the stem as it is transported to the side of the stem with more amyloplasts. Therefore the cells on the underside elongate more so that the stem grows upwards.

b A is the control because this removes the effects of gravity. It shows that the result in B is due to the effect of gravity, and no other factor, because the position of gravity is the only factor that is different between the two investigations.

20 a Woodlice tend to move towards dark, damp conditions. (Light, dry conditions are least favourable)

b Observe the position of each woodlouse every (say) 30 seconds, to determine the speed of movement and the rate of turning. If the woodlouse moves more-or-less straight towards the dark, damp conditions this is a taxis but if the rate of movement and rate of turning is faster in light, dry conditions but slower in dark and/or damp conditions, then this is a kinesis.

21 This means that the person is aware that the object is hot and can make a conscious decision not to touch it again. If this didn't happen, it would be possible that the person might try to touch the hot object again.

22 These animals are active only when there is little light present, so the light intensity is rarely enough to stimulate cones. However, the advantage of having mostly rod cells is that they are sensitive to low light intensity and the animals can see well in these conditions.

23 Blood returns to the lungs at a faster rate, so more carbon dioxide is expired.

24 Carbon dioxide concentration stimulates the Bohr effect in which the oxyhaemoglobin dissociation curve shifts to the right. This reduces the affinity of haemoglobin for oxygen, releasing more oxygen to actively respiring tissues.

Knowledge check answers

25 Note that the cardiac output and stroke volume are given in different units, so you need to convert stroke volume to dm^3 before calculating the heart rate. $210\,cm^3$ is $0.210\,dm^3$.

Rearranging the equation, $R = \dfrac{CO}{V}$

$= \dfrac{38}{0.210}$

$= 181$ beats per minute.

26 When an athlete exercises, the stroke volume increases. At rest, an athlete needs about the same cardiac output as a non-athlete. However, as their stroke volume is greater, the heart rate will be lower.

27 If the sodium ions diffuse 'backwards', the sodium channels will not open because they are in their refractory period. It is only the sodium channels 'in front' that can open.

28 Active transport moves molecules against their concentration gradient, while facilitated diffusion moves molecules down their concentration gradient; active transport uses energy from ATP, whereas facilitated diffusion does not require additional energy.

29 Acetylcholine is only produced in the presynaptic neurone, and its receptors are only found in the postsynaptic neurone.

30 To actively transport calcium ions back out of the cell; to actively transport acetyl and choline back into the cell; to re-synthesise acetylcholine.

31 Acetylcholine would not be broken down, so it would remain in the receptor sites and cause repeated action potentials in the postsynaptic neurone.

32 It stops sodium ions entering the postsynaptic neurone, so impulses from the sensory receptors cannot reach the central nervous system. Therefore no pain can be felt.

33 a Narrower
  b Stays the same
  c Narrower

34 Hydrolysis

35 Disulfide bridges are strong covalent bonds so they are harder to break. Ionic and hydrogen bonds depend on weak charges. A change in pH can change the charges, so the bonds break. Covalent bonds are less likely to break. Increased temperature gives the molecule more kinetic energy, so the weak bonds are more likely to break than the stronger covalent bonds.

36 Heart rate increases when the chemoreceptors detect that blood pH is too low. This results in more carbon dioxide being lost from the lungs and pH rising.

37 This allows blood glucose concentration to be controlled more reliably, since there is one hormone to reduce blood glucose concentration and another to raise it.

38 Insoluble, so it doesn't affect osmosis; compact, so a lot can be stored in a small space; branched, so there are lots of 'ends' to release glucose from.

39 They would be told to exercise regularly, especially after meals, so that glucose is metabolised faster. They would be advised to have small, frequent meals that are low in sugar but high in fibre and complex polysaccharides such as starch, so that blood glucose concentration increases slowly.

40 It is a countercurrent system because the filtrate in the descending part of the loop of Henle in the nephron flows in the opposite direction to the filtrate in the ascending limb. Therefore a concentration gradient is maintained all the way along the loop of Henle.

41 The more oxytocin is released, the more intense and frequent the contractions. This increases the concentration of oxytocin and therefore the frequency and intensity of contractions. In other words, the stimulus leads to a response that increases the change from the normal level.

Index